The Mysterious Patents Of
John Quincy St. Clair

The Illustrated Edition

Volume 1

Patents By:

John Quincy St. Clair

Compiled and Edited By:

Sebastian Aguanno Jr.

Index

Chapter 1: Internet Cellular Phone Prepaid Service

Abstract

A computer utility website service that allows cellular phone cards to be purchased over the Internet.

Classifications

G06Q20/18 Payment architectures involving self-service terminals [SSTs], vending machines, kiosks or multimedia terminals

US20050287984A1
United States

Inventor
John St.Clair

Worldwide applications
2004 US

Application US10/878,199 events

2004-06-28

Application filed by St Clair John Q

2004-06-28

Priority to US10/878,199

2005-12-29

Publication of US20050287984A1

Status

Abandoned

Description

BRIEF DESCRIPTION OF THE INVENTION

- [0001]

This invention is a utility service that allows cellular phone cards to be purchased over the Internet.

BACKGROUND OF THE INVENTION

- [0002]

At the present time there are only two ways that prepaid cellular phone cards can be purchased. The first method is to purchase them at a convenience store or gasoline station that sells the prepaid cards. The amount of the card is preprinted on the front of the card. The prepaid number is printed on the back of the card under a scratch-off pad. Using a special access code on the phone, such as *77 for Cingular wireless, the number is entered into the cellular phone in order to credit the user's account with the number of calling minutes corresponding to the amount paid.

- [0003]

The second method is to purchase the cards at a supermarket or similar outlet store that has a kiosk machine connected by modem to the store's dial-up telephone. The kiosk has a numeric keypad and liquid crystal display (LCD) for selecting the prepaid card desired from photographs displayed on the front panel. After inserting the appropriate amount of cash into the dollar bill reader, the kiosk prints out a paper receipt with the pin card number. The process proceeds as in the first method. The kiosk has a microcontroller which verifies that the amount paid is correct. It then dials the cellular phone company to obtain a proper pin number from their server computer.

- [0004]

There are a number of problems associated with the abovementioned methods which the inventor has personally experienced. First of all, the store may not be open which makes it impossible to obtain a card. If the store is open, they have run out of cards to purchase. If the store is open, cash might not be available for the kiosk machine because the ATM machine is out-of-order. In neighborhoods where crime is a problem, one does not feel safe parking the car on a strange side street and getting out of the car in order to walk to the store. Lastly, a store might not be available along a desolate country highway.

- [0005]

Maintenance is a problem with the kiosk machines. One problem is that the printer runs out of paper. Sometimes the store employees inadvertently disconnect the cable modem line when moving the phone. In other cases, the cleaning people disconnect the machine in order to plug in their vacuum cleaners, and then don't bother or remember to reconnect it.

- [0006]

In the last few months, more wireless technology has become available in the form of Internet-accessible portable data assistants (PDA) and wireless notebook computers. These devices allow one to connect to the Internet at any time of day or night without having to leave the car or house. Having a PDA means that the user can access a website and purchase a cellular phone pin number using a credit card, or funds from a prepaid account. Once the transaction is approved, the website server transmits the pin number over the Internet to the PDA which allows the phone caller, in a convenient and safe manner, to obtain more calling minutes.

SUMMARY OF THE INVENTION

- [0007]

It is the object of this invention to create a new method for purchasing cellular phone card pin registration numbers over the Internet using wireless PDAs and portable notebook computers. The purpose of the pin registration number is to purchase more call time on the cellular phone from the cellular phone company. The call time and service fee is paid for using a credit card or prepaid funds.

- [0008]

The invention consists of a website computer server connected to the Internet by either a dedicated subscriber line (DSL) or wireless satellite antenna transmission (SAT) system. The DSL transmits information (voice, video, data) at high speed over existing copper telephone lines.

- [0009]

The main server program is programmed in Sun Micro Systems object-oriented JAVA Internet programming language. The program initiates a JAVA listener which creates a new thread for each new user accessing the website. This makes the program multi-threaded which allows more than one user to be using the system at one time. Each thread displays a convenient Windows-like environment (check boxes, pull-down menus, buttons) which allows the user to create a new membership account, or log-on to purchase a pin number.

- [0010]

If the user is new to the system, the program allows the user to set up a new account with full name and verified password together with the user's address, telephone number and e-mail address.

- [0011]

If a previously registered user is logging onto the system, the program asks for the user name and password. The program then does a database lookup to verify this information. If the information is correct, the program requests the name of the cellular phone company from a pull-down menu containing a list of companies. Two check boxes allow the user to select a payment method by credit card or with prepaid account funds.

- [0012]

If payment is by credit card, the program displays a pull-down menu containing a list of credit card companies, a text box for entering the credit card number, a pull-down menu for the month of expiration and a pull-down menu for the year of expiration.

- [0013]

  If payment is by prepaid account funds, the program does a database look-up to read the account balance in the member's account. The program then displays this amount on the screen so that the user will know what level card to purchase.

- [0014]

  The program then displays a pull-down menu showing the various prepaid card amounts from which to select. An error message is displayed if the card amount is more than the funds available. An optional feature is that the user is given a minimum credit amount with which to purchase time in case of some emergency. Otherwise, the program displays the card amount plus the service fee amount to calculate the total charge to the credit card or account. At this point the user can cancel or continue with the transaction.

- [0015]

  If the user cancels, the program ends the JAVA listener thread and logs the user out. If the user continues with the transaction, the program connects to the credit card company to verify the transaction. If verified, the program connects to the cellular phone server to obtain the pin number which is displayed on the screen.

- [0016]

  The novelty of this system is that the website company can purchase blocks of prepaid cards whose pin registration numbers are entered into the server's database. It is occurring more frequently that the cellular phone company's system is unavailable in which case it is not possible to obtain a pin number immediately. What this means is that if the credit card transaction is approved, and the cellular phone connection is unavailable, the program can still provide a pin number to the user from the prepaid inventory. Thus the system is always 100% successful in obtaining a pin registration number.

  A BRIEF DESCRIPTION OF THE DRAWINGS

- [0017]

  FIG. 1. New membership window.

- [0018]

  FIG. 2. Log-on returning membership window and payment options.

- [0019]

  FIG. 3. Transaction approved window showing prepaid card number.

  DETAILED DESCRIPTION OF THE INVENTION

- [0020]

  1. The new membership window is shown in FIG. 1. It contains the user's full name and mailing address together with the telephone number and e-mail address. The user enters the password and then re-enters it a second time to verify the first entry. This information is then written to the customer master file in the Oracle database server with the name and password acting as the access keyword for record look-up.

- [0021]

  2. The returning membership window is shown in FIG. 2. The user enters the keyword consisting of the full name and password which allows the program to read the database customer master record for the phone number as well as the customer transaction file for the account balance. After the user enters the phone number, the program double checks the user's identity by comparing it to the phone number stored in the customer master file record.

- [0022]

  3. The user then selects the method of payment by clicking on the check boxes for either pay by credit card or pay with account funds. If the method is by credit card, the user enters the name of the credit card company using a pull-down menu, the credit card number, the expiration month by pull-down menu, and the expiration year by pull-down menu (shown by the black inverted triangle).

- [0023]

  4. If the user selects pay with account funds, the program reads through the database transaction file in order to obtain the account balance. This amount is then displayed on the screen in the Funds Available field.

- [0024]

  5. The user then selects the appropriate prepaid card from a pull-down menu. These cards come in various denominations such as $10, $15 and $20. The program then tallies the card amount with the service fee and then displays the total charge.

- [0025]

  6. The user then has the option of continuing with the transaction by clicking on the Accept button, or by clicking on the Cancel button. If the Cancel button is chosen, the program aborts the JAVA thread and logs out the user. If the Accept button is selected, the program proceeds to the next transaction approved screen.

- [0026]

  7. The Transaction Approved screen is shown in FIG. 3. This screen verifies that the transaction has been approved by displaying the prepaid card number. As an extra precaution, the prepaid number is sent to the user's e-mail address. The user can then enter this number directly into the cellular phone, or access the e-mail notification. The user then logs out by clicking on the Exit button which completes the transaction.

Claims

1. An Internet Cellular Phone Prepaid Service comprising:

a. A computer server connected to the Internet by a DSL or SAT line for the purpose of listening and responding to user requests to purchase prepaid phone calls by means of wireless portable data assistants (PDA), notebook computers and other types of Internet-connected data communications equipment such as home desktop computers;

b. A database server program written in object-oriented JAVA with multithreading capability that implements an Internet listener which allows the user to log onto the server and purchase a prepaid phone card by using a credit card or by prepaid account funds

c. An Internet communications and protocol program running on the server for connecting to the credit card company in order to verify the credit card information such as the card number and expiration date;

d. An Internet communications and protocol program running on the server for connecting to the cellular phone company in order to obtain a prepaid phone card number;

e. A database system for storing blocks of prepaid phone card numbers in case the cellular phone company's computer server is down;

f. A means by using a Windows-like environment (buttons, pull-down menus, check boxes, text entry fields) to display to the user the number of the prepaid phone card with which to obtain from the cellular phone company more calling minutes corresponding to the amount of the prepaid card.

g. An Internet communications and protocol program running on the server for e-mailing the user's account with the prepaid card number in case the user cannot record the number manually after the information is displayed on the screen.

NEW MEMBERSHIP

Full Name: _____

Password: _____

Verify Password: _____

E-mail Address: _____

POSTAL ADDRESS

Address Line1: _____

Address Line2: _____

City: _____

State/Region: _____

Zip Code: _____

Country: _____

Phone Number: _____

Figure 1

NEW MEMBERSHIP

Full Name: _____

Password: _____

Verify Password: _____

E-mail Address: _____

POSTAL ADDRESS

Address Line1: _____

Address Line2: _____

City: _____

State/Region: _____

Zip Code: _____

Country: _____

Phone Number: _____

Figure 2

RETURNING CUSTOMER

Full Name: _____

Password: _____

Phone Number: _____

PAYMENT METHOD

☐ **Pay by credit card**

Credit Card Company: _____ ▼

Credit Card Number: _____

Expiration Month: _____ ▼

Expiration Year: _____ ▼

_____

☐ **Pay with account funds**

Funds Available: _____

_____

Card Selection: _____ ▼

Card Amount: _____

Service Fee: _____

Total Amount: _____ | Accept | | Cancel |

Figure 3

TRANSACTION APPROVED.

Your prepaid card number is: _____

Your prepaid number has been sent to your e-mail address.

Thank you for using Internet Cellular Phone Prepaid Service.

Exit

Chapter 2: Walking Through Walls Training System

Abstract

This invention is a training system which enables a human being to acquire sufficient hyperspace energy in order to pull the body out of dimension so that the person can walk through solid objects such as wooden doors.

Classifications

G09B19/00 Teaching not covered by other main groups of this subclass

US20060014125A1
United States

Inventor
> John St. Clair

Worldwide applications
2004 US

Application US10/890,635 events

2004-07-14

Application filed by St Clair John Q

2004-07-14

Priority to US10/890,635

2006-01-19

Publication of US20060014125A1

Status

Abandoned

Description

BRIEF SUMMARY OF THE INVENTION

- [0001]

  This invention is a training system that enables a human being to acquire sufficient hyperspace energy in order to pull the body out of dimension so that the person can walk through solid objects such as wooden doors.

  BACKGROUND OF THE INVENTION

- [0002]

  A human being is a hyperspace energy being living in a physical container or body that is comprised of 67% water. This high percentage of water makes this invention possible. Referring to FIG. 1, the hyperspace energy being receives energy from our dimension through seven vortices that run the length of the body. Each vortex connects to a separate hyperspace dimension having its own particular frequency. This arrangement allows for the development of seven modular energy components corresponding to the mind, spiritual eye, voice, body, abdomen, plasma energy ball (battery), and ground connection.

- [0003]

  Vortex (A), known as the top vortex, supplies energy to the mind and provides a channel of communication to other entities in the universe. This channel has been tested up to 100,000 light years which is the diameter of the galaxy.

- [0004]

  Vortex (B), known as the spiritual eye, has a hollow cone-like appearance surrounded by white, misty, low-density hyperspace energy. Because the eye is modular, the mind can project it to vast distances, a process known as remote viewing.

- [0005]

  Vortex (C), known as the voice and hearing module, can also be projected to enormous distances in order to communicate verbally with other entities. Using the proper remote viewing instruments, it is possible to project both the eye and verbal modules to see and talk at the same time.

- [0006]

  Vortex (D), known as the heart vortex, provides protective energy to the upper internal organs, arms and hands.

- [0007]

  Vortex (E) provides protective energy for the lower organs such as the intestines, liver and kidneys.

- [0008]

  Vortex (F), known as the battery of the body, consists of an orange-colored plasma energy ball about one to two inches in diameter. The body becomes paralyzed if this ball is removed from the body. When the hyperspace energy being leaves the body, the vortices close and the battery takes over in order to keep the body functioning. This vortex also plays a role in the creation of the astral energy baby that attaches to the fertilized egg.

- [0009]

  Vortex (G) is the ground vortex which rotates in the counter-clockwise direction in order to provide an energy ground for the electrical circuit. All the other vortices rotate in the clockwise direction as seen from the front such that energy flows into the vortex according to the right-hand rule of physics.

- [0010]

  Vortex (H) is the hand vortex which rotates counterclockwise on the right hand and clockwise on the left hand as seen looking at the palm. Thus there is a rotating flow of hyperspace energy between the two hands when the palms are facing each other.

- [0011]

  Referring to FIG. 2, each vortex feeds energy into its own hyperspace module shown by the lettered box. In terms of quantum mechanics, each box is actually a potential energy well in which each module develops separately. At the time of death of the body, these modules are joined together as a single energy being. The process is powered by the plasma energy ball battery which also contains the logical instructions for assembly.

- [0012]

  Different people, as tested by the pendulum on the hand vortex, have different vortex sizes. Three people were tested. The first person had almost no rotational movement showing very little energy. The second person had a vortex radius of one inch. Another had a vortex radius of four inches which covered his entire hand. The latter also has the ability to lift another human being, lying prone on a table, off the table by flowing low-density hyperspace energy into the person through the hand vortices. He has also experienced walking through a solid wooden door with a dog at his side. What this means is that there is a way, as described in this invention, of creating large energy vortices which will enable a person to acquire sufficient energy to walk out of dimension through solid objects such as wooden doors.

- [0013]

  Researching the historical records, referring to the statue in FIG. 3, a humble black Catholic Dominican friar of the Santo Rosario Convent, by the name of San Martin de Porres, living in Peru in the 1500's, also developed this ability to walk through doors. For his beatification, many witnesses came forward to recount his extraordinary abilities. For example, a witness, who worked in the Convent, went to the cell of San Martin to ask for something to eat. When he reached the cell, he saw San Martin leaving with some medications apparently to heal someone sick. The witness waited by the open door of the cell for his return. After having waited awhile without being distracted by anything else, he saw said venerable brother fray Martin come out from his cell from the inside and call him by name. The witness was terrified, not understanding how this was possible.

- [0014]

The woman who gave me the statue told me that he would walk miles and miles each day to visit the poor. The task of walking means that there is a velocity involved. Because the body has mass, then there is a mass times a velocity, or momentum, involved in this ability. Notice also that the statue shows him walking with his right arm crossed over his left arm in the form of an X.

- [0015]

What this walking momentum means in terms of physics is given in the following analysis. The human body consists of 67% water. A water molecule consists of two hydrogen atoms and one oxygen atom having the atomic formula H_2O. The atomic weight of one atom of hydrogen is 1.008 awu. The atomic weight of one atom of oxygen is 16.000 awu. The molecular weight of one atom of water is therefore: Weight of two atoms of hydrogen 2 × 1.008 awu = 2.016 awu☐Weight of one atom of oxygen 1 × 16.000 awu = 16.000 awu 18.016 awu

The formula weight is just the atomic weight expressed in grams. Thus the formula weight of water would be 18.016 grams or 0.018016 kilograms. According to Avogadro's law, the formula weight contains $N=6.02 \times 10^{23}$ molecules. Thus the mass of one water molecule is the formula weight divided by the number of molecules: mass = .018016 ☐ ☐ ☐ kg N = 2.99269103 · 10 - 26 ☐ ☐ ☐ kg ln ☒ (mass 1 ☐ ☐ ☐ kg) = - 58.77103943

- [0016]

According to Einstein's Special Theory of Relativity, energy is equal to the mass times the speed of light squared. The energy of a photon is equal to Planck's constant h times the frequency f of the photon. Equating these two energies shows that E = mc 2 ≈ hf = h ☐ c λ = h c ☐ c 2 λ ☐ ⇒ m ☐ ☐

☐ λ = h c ☐ ⇒ ln ☒ (m) + ln ☒ (λ) = ln ☒ (h c) = base = - 95.91546344 which says that the natural logarithm of the mass plus the natural logarithm of the wavelength is equal to the natural logarithm of Planck's constant divided by the speed of light c, known as the base constant in the tetrahedron diagram. This diagram plots the mass versus wavelength in natural logarithms. Notice that the left hand side of the equation is the sum of mass and wavelength, so the right hand side must also be the same. ln ☒ (h c) = ln ☒ (Ω∧ ☐ ☐ ☐ 2 ☐ π∧ ☐ ☐ ☐ c c) = ln ☒ (Ω∧) + ln ☒ (2 ☐ π∧) where Ω∧, known as the Planck mass, is the linear mass Ω of the universe times the bottom dimensional limit of the universe ∧, and 2π∧ is the bottom dimensional wavelength, known as the Planck wavelength. That is, our dimension is bounded by the Planck box having sides Planck mass and Planck wavelength. These boundaries have values:
ln(Ω∧)=−17.64290101
ln(2π∧)=−78.27256243

- [0017]

Referring to FIG. 4, these two lines are plotted on the tetrahedron diagram. The Planck mass line (A) reflects off the sphere (C) and returns as the Planck wavelength (B) which shows the dual nature of quantum physics. This creates the Planck box (a,b,c,d) which is the boundary of our dimension.

- [0018]

Referring to FIG. 5, the mass of the water molecule is plotted as horizontal line (D) on the diagram. The energy of the water molecule is the mass times the speed of light squared.

$E=mc^2$

$\ln(c)=19.51860099$

$\ln(mc^2)=\ln(m)+2\ln(c)$

- [0019]

  Referring to FIG. 6, a circle (E), having a radius equal to the speed of light squared, centered on the mass of the water molecule at the vertical axis, generates a circle (F), centered at the origin, that intersects (e) the mass of the water molecule at the Planck wavelength. That is, this intersection point sits right on the Planck box boundary between space and hyperspace. The radius of circle (F) is actually the mass of the water molecule divided by the speed of light squared. $E = mc^2 = m \ ⊠ \Rightarrow 1c2 = 1 \ ⊠ \Rightarrow c = 1 \ ⊠$ meter sec ln ⊠ (c) = 0 Taking the positive square root, the speed of light is one meter per second at the Planck boundary. The experiments with brain hemisphere resonance show that the resonant frequency of the human energy field is between 1 Hz to 5 Hz which is well below the hearing threshold of 20 Hz. Because the traveling wave has a wavelength of 0.3048 m and the speed of light is unity at the boundary, the frequency should be $f = c \lambda = 1 ⊠$ m sec .3048 ⊠ ⊠ ⊠ m = 3.28 ⊠ ⊠ ⊠ Hz which is within the middle of the experimental male range. At this resonant frequency, the human energy being pops out of the body. This represents only a first stage in the development of the energy being. But what is really wanted is to have both the physical body and the energy field move out of dimension together as San Martin did.

- [0020]

  Referring to FIG. 7, the 45° base line (G) is added to the diagram. Notice that the Planck mass intersects (b,d) the Planck wavelength on this line because they sum to the base constant. The mass of the water molecule crosses this line at point (f). A circle, centered (f), with a radius equal to the speed of light, is tangent to the Planck wavelength (h) and the Planck mass (g). Since mass times velocity is momentum, the diagram says that the momentum of the water molecule is tangent to the boundaries of the Planck box which separates space from hyperspace. In order to get to point (e) from the momentum of the water molecule, a second circle is added to the momentum.

- [0021]

  Referring to FIG. 8, a circle (I), centered on the water molecule mass at the speed of light circle (i), is made tangent to the Planck wavelength at point (e). The momentum M of space is equal to the Planck mass times the Planck scale times the speed of light. At point (e), the speed of light is unity, so that the momentum is just the Planck mass in momentum units: M = Ω∧c = 2.176634194 · 10 - 8 ⊠ ⊠ ⊠ kg ⊠ m s
  Circle (I) has a radius equal to
  ratio=$e^{21.572952}$=2338912700
  Therefore the walking momentum in order to get to point (e) is the momentum of space M times this ratio M w = M · ratio = 50.909573606 ⊠ ⊠ ⊠ kg ⊠ m sec
  The stride length L per second that a person of mass W has to walk is the walking momentum divided by the mass W times a period T of one second L = M w W ⊠ T
  For a person with a mass W of 99.79 kg (220 pounds), then the stride length L is 20.08 inches or one foot and eight inches. The person has to walk this length in one second on each foot.

- [0022]

  Looking at the statue of San Martin, his arms are crossed over each other. The vortex of the right hand points backward, and the left hand vortex points forward due to the reversed rotation. Referring to FIG. 9, this creates a rotational energy channel (D) around his body (A). The stride length (C) is calculated according to the body mass, and then a banner printout is made showing where the footprints (B) are to be placed each second. The question is: "What happens when one walks the walk?".

- [0023]

  On the very first experiment, referring to FIG. 10, what happens is that, after taking only six strides on the banner printout (A), a huge spinning vortex (C) develops over the top of the head and the vertex locks onto the heart vortex in the center of the chest (B). In everyday life, this vortex is not created because normal walking is much faster and the hands are held at the side of the body. The energy rush through the pineal gland is so intense that one feels immediately sleepy and starts yawning excessively due to the increased flow of melatonin.

- [0024]

  After practicing with the banner printout, long walks were made through the park. In this case, a vertical white line rotated around a vertical axis located about six feet perpendicular to the path on the right side of the body. When the walking speed was correct, this white line would lock onto the centerline of the body. Speeding up or down caused the white line to lose synchronization and rotate away. This white line is related to the ability to levitate the body. San Martin had so much energy that, according to witness testimony, he could float horizontally in the air with his head resting against the bowed head of Christ on a carved wooden cross. Thus San Martin's energy sources were channeling energy from Christ, collective broom energy as described in a separate patent application, and the walking momentum vortex energy.

- [0025]

  During the early part of the 20th century, a man's parents were lying in bed dying of tuberculosis. With their permission, he placed a weighing scale under each of their beds. When they passed away, he found that each scale registered a loss in weight of 2.5 ounces. This is equivalent to 0.071 kg, which is the mass of the human energy being.

- [0026]

  After conducting a number of experiments with water vortices draining from a cylindrical tank, it can be stated from Bernoulli's theorem that the potential energy plus the kinetic energy is a constant $gz + \frac{1}{2} mv^2 = const$
  The shape of the inner surface of the water circulation has a velocity proportional to the inverse of the radius, so the shape of the surface is $(z - z_0) = \frac{k}{r^2}$
  which says that the height of the vortex is proportional to the inverse of the square of the radius.

- [0027]

The hand vortex area ratio between the second test subject and the third test subject is equal to the square of their radii: ratio = $(1 \square \square \square in)^2 (4 \square \square \square in)^2 = 1\ 16$

Because the speed of light at the boundary was determined to be one meter per second, the energy of the third test subject is $E = (.071 \square \square \square kg\ 16) \square (1 \square \square \square m\ sec)^2 = 4.4375 \cdot 10 - 3 \square \square \square joule\ ln \boxed{} (E) = - 5.417664124$

- [0028]

Referring to FIG. 11, a circle (K), having this radius, is added to the energy of the water molecule (E), to produce augmented energy circle (J). This circle (J) intersects the mass of the water molecule outside the Planck box at point (j). This means that the increased hyperspace energy moves the water molecule, and hence the body, out of dimension. Furthermore, circle (J) is tangent to the walking momentum ratio circle (I) which keeps the geometry locked together.

SUMMARY OF THE INVENTION

- [0029]

It is the object of this invention to create a training system that allows a person to develop the ability to walk around out of dimension, passing through solid objects. This invention is based on one of the most remarkable relationships between the water molecule and the boundary between space and hyperspace. The mass of the water molecule is equal to the energy of the water molecule at this boundary. Because the body is composed of 67% water, the body sits on the boundary such that any additional increase in energy would move the body out of dimension into hyperspace. Because human beings are actually hyperspace energy beings living in physical bodies, the additional energy required to move the body out of dimension comes from increasing the energy of the hyperspace being. One source of this energy comes from walking cross-handed at the proper velocity in order to generate a large hyperspace energy vortex that flows energy into the potential wells of the hyperspace being. This increased hyperspace energy will then allow the person to walk around out of dimension through solid wooden doors. Because the door and the person are in two slightly different dimensions at the same moment, it appears that the person is walking through the door. After passing through the closed door, the person then returns to our dimension and emerges in the interior of the closed-door room.

- [0030]

This technique can be used in reverse to heal an infected hand instantaneously. A salve made from the St.Mary's herb is applied to the skin of the infected hand. The hyperspace energy then flows through the right-hand vortex such that the infected hand and the salve are taken slightly out of dimension. What happens is similar to when a short piece of straw is embedded in a hard wooden telephone pole as a tornado passes over the pole. The straw and pole are taken out of dimension such that they briefly merge together. As the tornado moves on, both objects come back to dimension merged together. Thus the salve (straw) is merged with the bacteria (pole) in hyperspace such that the bacteria is killed instantly. Removing the hand vortex brings the infected hand back into dimension cured.

- [0031]

Based on this information and the results of many experiments, this invention creates a large vortex by walking at a certain velocity with the hands crossed over the chest. The proper walking momentum is created by a computer program that inputs the person's weight, shoe length, and the number of strides to be taken. The program then prints out a banner showing the footprints where the person has to step each second. When a person obtains sufficient energy from these methods, the person is then tuned to the subspace geometry of the universe as will be shown using the tetrahedron physics diagram.

A BRIEF DESCRIPTION OF THE DRAWINGS

- [0032]

  FIG. 1. Energy vortices of the human body.

- [0033]

  FIG. 2. Seven potential wells fed by the energy vortices of the body.

- [0034]

  FIG. 3. Carved wooden statue of San Martin de Porres who could walk through solid wooden doors.

- [0035]

  FIG. 4. Tetrahedron diagram showing boundaries of the Planck box of dimension.

- [0036]

  FIG. 5. Tetrahedron diagram showing mass of water molecule.

- [0037]

  FIG. 6. Tetrahedron diagram showing water molecule energy and mass are equal at the Planck box boundary.

- [0038]

  FIG. 7. Tetrahedron diagram showing that water molecule momentum is tangent to the boundaries of the Planck box.

- [0039]

  FIG. 8. Tetrahedron diagram showing the momentum ratio required to reach the Planck wavelength boundary from the water momentum.

- [0040]

  FIG. 9. Perspective view of crossed-hand momentum walking using banner printout.

- [0041]

  FIG. 10. Perspective view of vortex generated by momentum walking.

- [0042]

  FIG. 11. Tetrahedron diagram showing how additional hyperspace energy supplied to the potential wells of the hyperspace energy being enables the human body to be pulled out of dimension.

- [0043]

  FIG. 12. Tetrahedron diagram showing the inverted tetrahedrons whose crossing represents the merging of two worlds between space and hyperspace.

- [0044]

  FIG. 13. Tetrahedron diagram showing that the proton wavelength is determined by the Planck mass tangent to the inverted tetrahedrons.

- [0045]

  FIG. 14. Tetrahedron diagram showing that the mass of the universe determines the electron and proton elementary particles.

- [0046]

  FIG. 15. Tetrahedron diagram showing that momentum walking together with the increased energy of the hyperspace energy being is tangent to the mass of the universe.

- [0047]

  FIG. 16. Computer program block diagram for printing banner footprints.

- [0048]

  FIG. 17. Computer program input dialog window.

- [0049]

  FIG. 18. Computer program calculation of stride length per second.

- [0050]

  FIG. 19. Six-stride screen banner printout for 220 lb. person.

- [0051]

  FIG. 20. Project tree showing help information by double clicking on node.
  DETAILED DESCRIPTION OF THE INVENTION

- [0052]

1. Referring to FIG. 12, a tetrahedron (A, path abc) is added to the diagram. The tip of the tetrahedron (e) falls on the base constant which is equal to Planck's constant divided by the speed of light. A second tetrahedron (B, path def) is inverted around horizontal line (D) which has a geometrical relationship to the base constant. The line is located at centerline = 2 3 ⟨⟩ ln ⟨⟩ (h c) = - 110.7536373
Notice that the intersection of the two inverted tetrahedrons (g) occurs at the Planck wavelength which is the boundary between space and hyperspace. Line (D) is referred to as the merging of two worlds or the connecting of two worlds, a phrase obtained by means of remote viewing. That is, it is the dividing line between space and hyperspace. This is the reason that the two boundaries intersect at this point.

- [0053]

    Referring to FIG. 13, the proton wavelength (E) is added to the diagram. The proton wavelength has a value of the electron wavelength divided by
    1836.1527 ln ⟨⟩ (λ p) = ln ⟨⟩ (λ e 1836.1527) = - 34.26005901
    A line (hd), from the base constant at the proton wavelength (h), to the corner of the inverted tetrahedron (d), intersects the merging of two worlds line at point (i). A circle, with a radius equal to the Planck mass (G), centered (i), is tangent to the inverted tetrahedrons. Thus the proton is defined by the base constant and the geometry of subspace. The reason that the proton is tangent to both tetrahedrons is because the electron and proton follow one single path between space and hyperspace. Thus there is only one single particle in nature. Because the particle enters our space at two different locations, we see the one particle as two distinct elementary particles. This relationship can be seen in Library of Congress tetrahedron diagram tet0565.

- [0054]

    Referring to FIG. 14, the mass of the universe MU is equal to the linear mass Q of the universe times the radius R of the universe which is 10^{26} meters
    $\ln(MU)=\ln(\Omega R)=122.3347509$
    as shown on the diagram as line (A). The electron wavelength (B) reflects off the circumscribing sphere (H) and returns as the electron mass (C). The distance between reflection points is the hyperspace charge which is equal to the charge of space less the electron charge. So the electron goes from wavelength to electric charge to mass. The proton wavelength (D) reflects off the sphere and returns as the proton mass (E) which intersects the horizontal axis at point (c). A line (abc) from the mass of the universe at the vertical axis (a) to the proton at the horizontal axis (c) intersects the electron (b) which determines the electron's mass and wavelength since this point is on the 45° base line. What this means is that cosmology determines the values of the elementary particles.

- [0055]

Referring to FIG. 15, the mass of the water molecule (C) intersects the 45° base line at point (a). A circle (F), with a radius equal to the Planck mass is centered on point (c) at the Planck wavelength boundary (B) where the mass of the water molecule numerically equals the energy of the water molecule and where the speed of light is unity. The Planck mass is tangent to the base constant which is the vertical centerline (I) of the diagram. The large circle (H), centered (a) on the water molecule, is tangent to the mass of the universe (b) and tangent to the inverted tetrahedron (E). The difference between this circle (H) and the base constant (I) is the energy the hyperspace being has to acquire in order to be tuned to the mass and geometry of subspace. This difference, shown as circle (G), has a mass $m = e^{-4.792671}$ kg = $8.29 \cdot 10^{-3}$ kg

Thus the hand vortex radius ratio has to be the square root of the mass of the hyperspace energy being divided by this tangent mass, or: r = .071 ⬚ ⬚ ⬚ kg · 1 ⬚ ⬚ ⬚ in 2 e - 4.792671 ⬚ ⬚ ⬚ kg ≈ 3 ⬚ ⬚ ⬚ inches

which is a vortex radius that is three times larger than that of a normal person, but one inch smaller than the third test subject who had a hand vortex radius of 4 inches. This is the reason that the third test subject was able to walk through walls and teleport to other locations because his energy was sufficiently large enough to cross over the inverted tetrahedron into a co-dimension of hyperspace. Notice also that dotted circle (J) with a radius equal to the mass of the water molecule, centered (c) on the boundary, is tangent to the tetrahedron (K). This makes the combined geometry tangent to the inverted tetrahedrons and the mass of the universe.

- [0056]

2. Referring to FIG. 16, a computer program generates a banner printout with footprints spaced for walking according to the weight, shoe size and length of banner desired. Some banners could fit in a small room, or be placed on the floor of a long corridor. As shown in the block diagram the program inputs these three variables with error checking. Then the program calculates the stride length L per second from the equation L = M W ⬚ T

- [0057]

3. In the above equation, the value of the momentum M, as determined by the tetrahedron diagram, is preferably 50.9095736 kg m/s. The weight of the person is converted to mass W in kilograms. The stride period is preferably 1 second. Referring to FIG. 17, the data is entered in the dialog input window.

- [0058]

4. Once the stride length has been calculated, the program displays the stride length and the required number of banner sheets in the message window, as seen in FIG. 18.

- [0059]

5. The print banner menu is selected and the foot prints are printed on continuous banner paper. A six-stride scaled screen version of the banner is shown in FIG. 19. The banner paper is then placed on the floor, and a one-second beeping timer is activated from the toolbar or menu. The person then walks beside the printed foot prints, taking one stride per beep, which produces the correct walking momentum to generate the hyperspace vortex. The vortex, which forms in only six strides, brings additional hyperspace energy into the quantum potential wells of the hyperspace energy being.

- [0060]

6. A help system consists of a project tree which explains the various steps in using the program. Double mouse clicking on a project tree node displays the help instructions in a dialog window as shown in FIG. 20.

- [0061]

7. In summary, the purpose of the training system is to substantially increase the energy of a human being who will then have the capability of walking through walls, body levitation, instantaneous healing of infections, full-body teleportation to another location, remote viewing at vast distances in terms of light-years, and looking into hyperspace co-dimensions. The third test subject and I have been able to experience all the above phenomena. He did it through augmenting his energy, and I have done it through the application of electromagnetic fields, by spinning on my vortex accelerator machine and using this invention.

Claims

1. A training method comprising the steps of:
generating a banner having a plurality of footprints spaced at regular intervals wherein the banner is placed on the ground; generating a periodic audible signal, whereby the audible signal repeats at a regular interval of time equal to the period; and
walking on the banner by tracing the footprints spaced at regular intervals, wherein one step is made with each audible signal.

2. The method of claim 1, wherein the step of generating the banner further comprises:
providing a person's actual mass in kilograms; and determining the stride length based upon the following equation:

$L = (M/W) * T$
where L is the stride length in meters, M is a constant of approximately 50.91, W is the mass of the person in kilograms, and T is the period of the audible signal in seconds, and wherein the footprints on the banner are spaced at the stride length.

3. The method of claim 2, wherein the audible signal is a beep and the period of the audible signal is one second.

4. A training method for a person comprising the steps of:
generating an audible signal having a fixed period;
generating a banner having regularly spaced indicia for identifying preferred step locations, wherein the distance between adjacent indicia is determined by the following formula:

$L = (M/W) * T$

where the distance between adjacent indicia in meters is L, a constant of 50.9095736 is equal to M, the person's mass in kilograms is W, and the fixed period in seconds is T; and walking on the marked path by stepping upon each of the regularly spaced indicia wherein one step is made with each period of the audible signal.

5. The training method of claim 4 wherein the fixed period is one second.

6. The training method of claim 5, wherein the regularly spaced indicia are footprints.

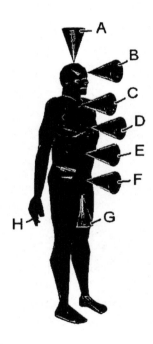

Figure 1

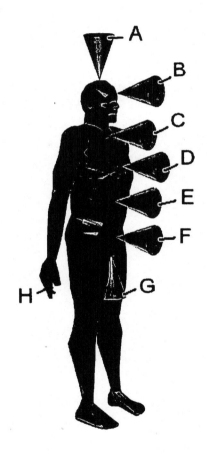

Figure 2

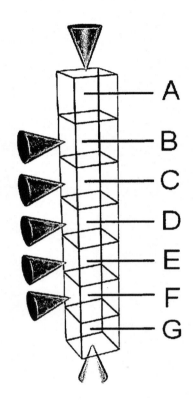

Figure 3

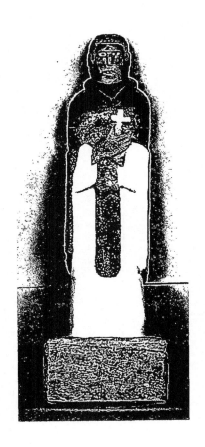

Figure 4

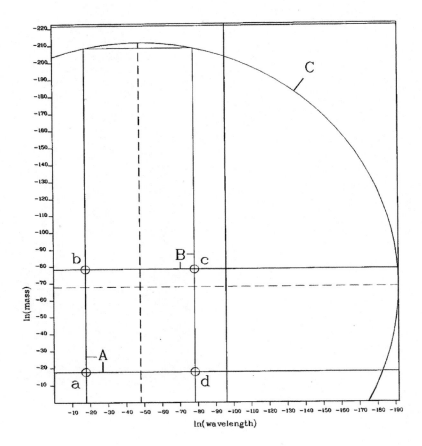

Figure 5

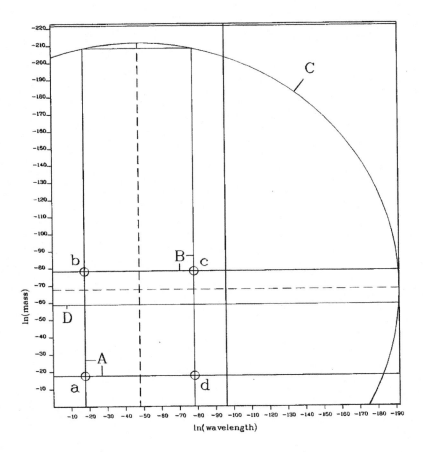

Figure 6

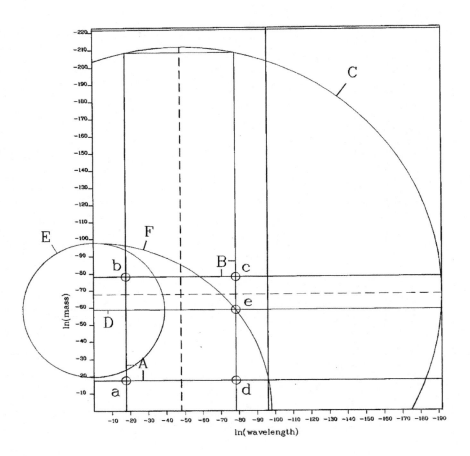

Figure 7

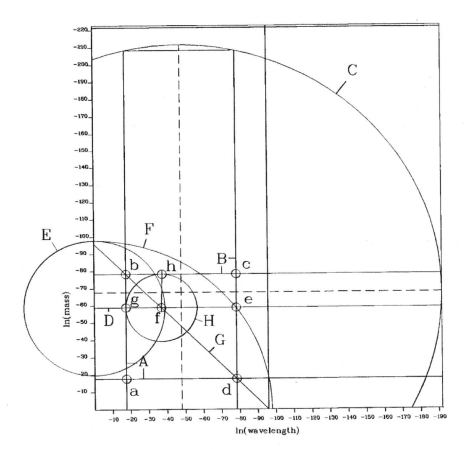

Figure 8

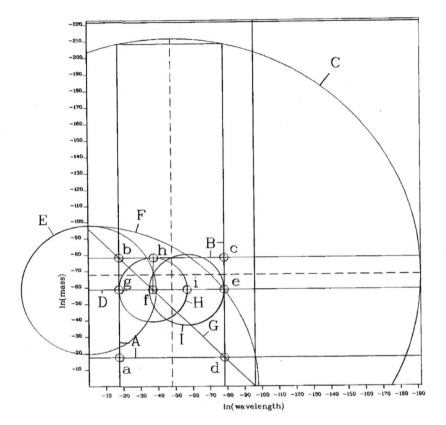

Figure 9

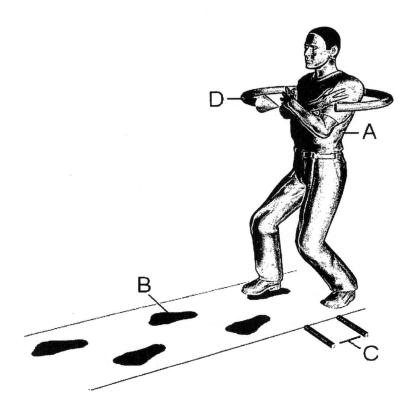

Figure 10

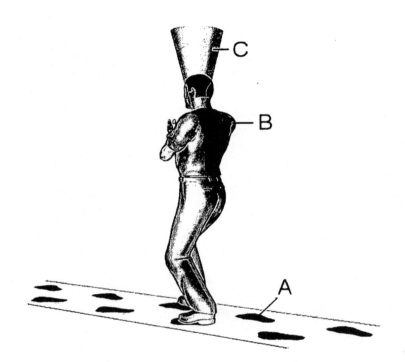

Figure 11

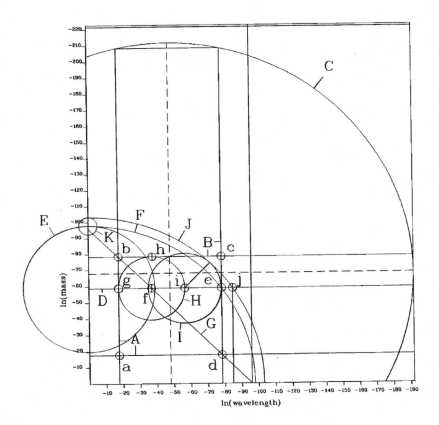

Figure 12

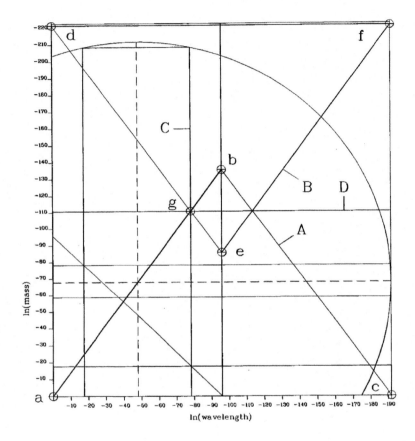

Figure 13

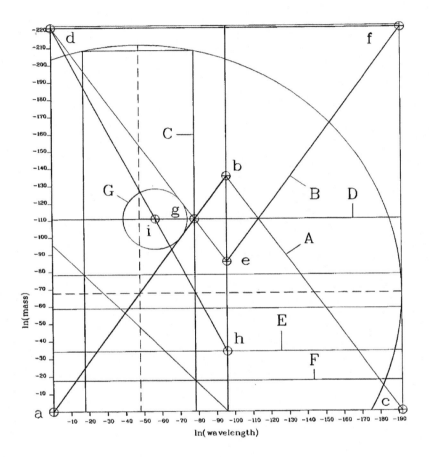

Figure 14

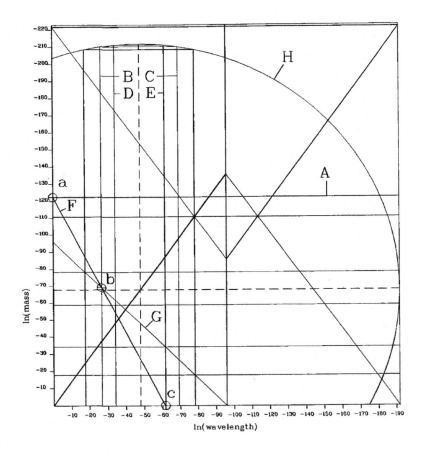

Figure 15

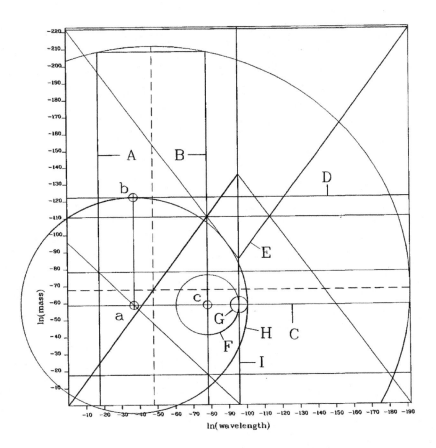

Figure 16

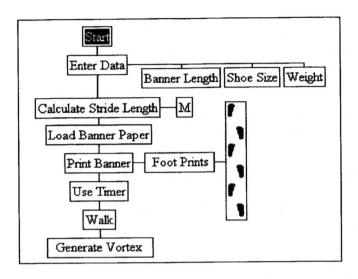

Figure 17

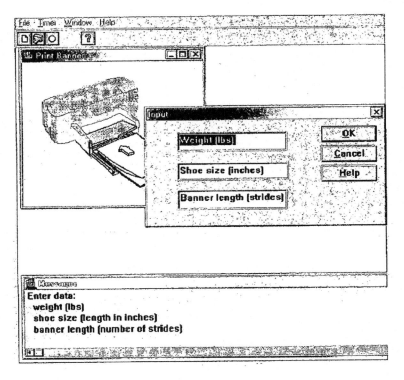

Figure 18

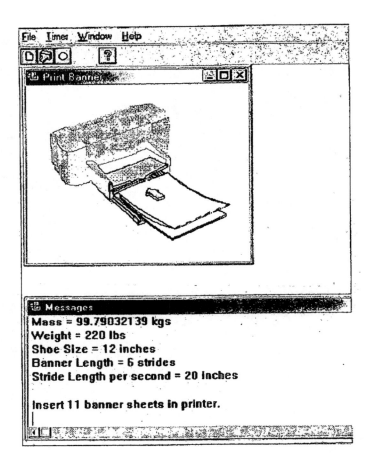

File Timer Window Help

Print Banner

Messages

Mass = 99.79032139 kgs
Weight = 220 lbs
Shoe Size = 12 inches
Banner Length = 6 strides
Stride Length per second = 20 inches

Insert 11 banner sheets in printer.

Figure 19

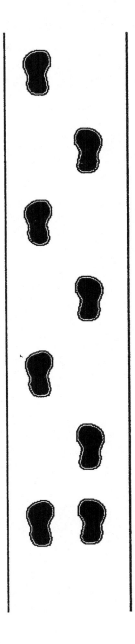

Figure 20

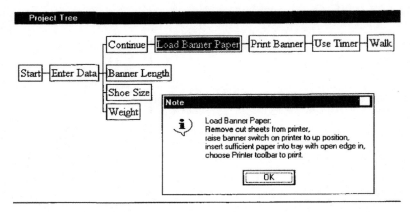

Chapter 3: Electric Dipole Spacecraft

Abstract

This invention is a rotating spacecraft that produces an electric dipole on four rotating spherical conducting domes perturbing a uniform spherical electric field to create a magnetic moment interacting with the gradient of a magnetic field that generates a lift force on the hull.

Classifications

B64G1/409 Unconventional spacecraft propulsion systems

US20060038081A1
United States

Inventor

John St. Clair

Worldwide applications
2004 US

Application US10/912,621 events

2004-08-04

Application filed by St Clair John Q

2004-08-04

Priority to US10/912,621

2006-02-23

Publication of US20060038081A1

Status

Abandoned

Description

BRIEF SUMMARY OF THE INVENTION

- [0001]

  This invention is a rotating spacecraft that utilizes four spherical conducting domes perturbing a uniform electric field in order to create a lift force by means of a magnetic moment times the gradient of a magnetic field.

BACKGROUND OF THE INVENTION

- [0002]

  An electric dipole p is two electrical charges of opposite sign {q, −q} separated by a distance a.
  $p=qa$=coulomb·meter

- [0003]

  If this dipole is moving with a velocity v, it produces a magnetic moment μ. μ = pv = qav = coulomb ⯑ ⯑
  ⯑ meter ⯑ meter sec = coulomb sec ⯑ meter 2 = IArea

- [0004]

  The magnetic moment is equal to a current I circulating around an area. The magnetic field B has units of kilogram per second per charge coulomb. B = kg sec ⯑ ⯑ ⯑ coul

- [0005]

  The gradient of the magnetic field in the vertical direction z has units of dB dz = kg sec ⯑ ⯑ ⯑ coul ⯑ ⯑ ⯑ meter

- [0006]

  This gradient interacting with a magnetic moment creates a force F measured in newtons. F z = μ ⯑ dB dz = coul ⯑ ⯑ ⯑ m 2 sec ⯑ kg sec ⯑ ⯑ ⯑ coul ⯑ ⯑
  ⯑ m = kg ⯑ ⯑ ⯑ m sec 2 = newton

- [0007]

  In terms of vectors, the force is equal to
  $F=\nabla(\mu \cdot B)$
  which is the gradient ∇ of the dot product (·) of the magnetic moment with the magnetic field. This means that the magnetic moment has to be aligned with the field. The lift force on the spacecraft would then be the magnetic moment in the vertical z-direction μ_z times the magnetic field in the z-direction B_z. For constant magnetic moment, the gradient affects the magnetic field only, resulting in the same force equation
  $F_z=\nabla(\mu_z \cdot B_z)=\mu_z \nabla B_z$

- [0008]

Referring to FIG. 1, the electric dipole has a positive charge q located on the z-axis at a distance a from the origin of the graph. A second negative charge −q is located at a distance −a from the origin. The positive charge produces an electrostatic potential φ_1 at a radius r_1 equal to the charge q divided by 4π times the permittivity of space ε_0 $\varphi 1 = q 4 \pi \varepsilon 0 1 r 1$

where the permittivity is linear capacitance, measured in farads per meter. The electrostatic potential has units of
volts $\varphi = coul (farad\ meter) 1\ meter = coul\ farad = volt$
because the charge in coulombs held by a capacitor is equal to the capacitance, measured in farads, times the capacitor voltage. Because the second charge has the opposite sign, the potential φ_2 at a radius r_2 to the same point in space is $\varphi 2 = - q 4 \pi \varepsilon 0 1 r 2$

- [0009]

  The total potential φ at some point in space is equal to the sum of the two potentials, or $\varphi = \varphi 1 + \varphi 2 = q 4 \pi \varepsilon 0 r 1 - q 4 \pi \varepsilon 0 r 2 = q 4 \pi \varepsilon 0 [1 r 1 - 1 r 2]$

- [0010]

  As seen in the diagram, the point of space is a distance r from the origin. Using the law of cosines, radius r_1 can be written as $r 1 = (r 2 + a 2 - 2 ar \cos (\theta)) 1 2 = r (1 + (a r) 2 - 2 (a r) \cos (\theta)) 1 2 = r (1 - 2 xt + t 2) 1 2$
  where t is the ratio of the charge location over the radius, and x is $\cos(\theta)$. The potential for positive charge q_1 can be written $\varphi 1 = q 4 \pi \varepsilon 0 1 r 1 = q 4 \pi \varepsilon 0 1 r (1 - 2 xt + t 2) - 1 / 2$

- [0011]

  Dropping the factor $q/4\pi\varepsilon_0 r$, the square root can be expressed in terms of the Legendre polynomial $P_n \cos(\theta)$ of the nth power $g (t , x) = (1 - 2 xt + t 2) - 1 / 2 = \sum n = 0 \infty P n (x) t n$
  where the absolute value of t is less than one. The polynomial coefficients of t^n can be obtained by using the binomial theorem to expand the generating function g(t,x) as $(1 - 2 xt + t 2) - 1 / 2 = \sum n = 0 \infty (2 n) ! 2 2 n (n !) 2 (2 xt - t 2) n$
  which evaluates to $0 ! 2 0 (0 !) 2 (2 xt - t 2) 0 + 2 ! 2 2 (1 !) 2 (2 xt - t 2) 1 + 4 ! 2 4 (2 !) 2 (2 xt - t 2) 2 1 t 0 + xt 1 + (3 2 x 2 - 1 2) t 2 + order (t 3)$

- [0012]

  The first three Legendre polynomials are therefore $P 0 = 1 \quad P 1 = x \quad P 2 = 1 2 (3 x 2 - 1)$

- [0013]

  The electrostatic potential for both charges of the electric dipole is $\varphi = q 4 \pi \varepsilon 0 1 r \{ (1 - 2 (a r) \cos (\theta) + (a r) 2) - 1 / 2 - (1 + 2 (a r) \cos (\theta) + (a r) 2) - 1 / 2 \}$

- [0014]

The potential can be evaluated in terms of the Legendre polynomials as $\varphi = \frac{q}{4\pi\varepsilon_0 r}\left[\sum_{n=0}^{\infty} P_n(\cos(\theta))\left(\frac{a}{r}\right)^n - \sum_{n=0}^{\infty} P_n(\cos(\theta))(-1)^n\left(\frac{a}{r}\right)^n\right] = 2\frac{q}{4\pi\varepsilon_0 r}\left[P_1(\cos(\theta))\left(\frac{a}{r}\right) + P_3(\cos(\theta))\left(\frac{a}{r}\right)^3 + \ldots\right]$

- [0015]

The first and most dominant term when the radius is much greater than location a is equal to $\varphi = \frac{2aq}{4\pi\varepsilon_0}\frac{P_1(\cos(\theta))}{r^2}$
which is the electric dipole potential and 2aq is the dipole moment
$p = 2aq$

- [0016]

Now imagine a constant electric field E_0 which is perturbed by a conducting sphere of radius a. The unperturbed electrostatic potential outside the sphere would the negative of the electric field times the radius times the Legendre polynomial, or
$\varphi_1 = -E_0 r P_1$

- [0017]

The electrostatic potential perturbed by the charges is the voltage E_0a times the radius times the a of the dipole moment times the Legendre polynomial divided by the radius squared $\varphi_2 = E_0 \frac{aaa P_1}{r^2} = E_0 \frac{a^3 P_1}{r^2}$

- [0018]

The total potential outside the sphere is the sum of the two potentials equal to $\varphi = -E_0 r P_1 + E_0 \frac{a^3 P_1}{r^2} = -E_0 P_1\left(r - \frac{a^3}{r^2}\right) = -E_0 r P_1\left(1 - \left(\frac{a}{r}\right)^3\right)$

- [0019]

Referring to FIG. 2, the previously uniform electric field is shown perturbed by the neutral conducting sphere. The center of the sphere is taken as the origin and the z-axis is oriented parallel to the original uniform field.

- [0020]

The electric field induces a surface charge density σ on the sphere equal to the negative of the permittivity of space times the gradient of the electrostatic potential $\sigma = -\varepsilon_0\left.\frac{\partial\varphi}{\partial r}\right|_{r=a} = 3\varepsilon_0 E_0 \cos(\theta)$

- [0021]

The electric field also induces an electric dipole moment on the sphere equal to the $p = qa \, r \cdot 2 \;\square\;\square\;\square\; a\, 2\, r = 4 \;\square\;\square\;\square\;\square\; \pi\varepsilon\, 0 \;\square\; E\, 0 \;\square\; a\, 3$

with units of coulomb-meter. If this sphere is rotating around a central axis at a velocity v, it will create a magnetic moment μ equal to the dipole moment times the velocity.

μ=pv

with units of ampere-meter^2.

SUMMARY OF THE INVENTION

- [0022]

    As shown in the preceding background section, a neutral conducting sphere placed in a uniform electric field will generate a magnetic moment when rotated around a central axis. The electric field can be created by two points charges of opposite sign separated by a distance between them.

- [0023]

    Referring to FIG. 3, the spacecraft has a spherical cabin (A) to which are attached cone-shaped electrostatic towers (B,C) above and below the cabin along the direction of travel in the z-direction. Because the electric field goes from the positive charge to the negative charge, the tip of the lower tower has a positively charged electrode, and the upper tower has a negatively charged electrode. Four equally-spaced neutral conducting spheres (D) are connected to the cabin by non-conducting tubes (E). The tubes make an angle with the cabin such that the distance (CD) is greater than distance (DB). The angle θ of the tube with respect to the cabin can be seen in side view FIG. 4.

- [0024]

    Referring to FIG. 5, the charges create a uniform spherical field between the towers. The conducting spheres perturb this field such that the electric field (E) points toward the upper tower in a manner similar to that shown previously in FIG. 2.

- [0025]

    Referring to FIG. 6, the conducting sphere produces an electric dipole moment (A) pointing at an angle toward the upper tower.

- [0026]

    Referring to FIG. 7, by the law of addition of vectors, the electric dipole {overscore (p)} can be represented by two orthogonal vectors pointing in the vertical z-direction p_z and in the inward radial direction p_r.

- [0027]

    Referring to FIG. 8, the hollow tube (A) connecting the cabin with the conducting sphere contains a spiral-wound electrical solenoid (B) which produces a magnetic field (C). This magnetic field {overscore (B)} can be decomposed into two orthogonal vectors pointing in the vertical z-direction B_z and in the outward radial direction B_r as shown in FIG. 9.

- [0028]

Referring to top-view FIG. 10, the spacecraft has a clockwise angular velocity ω (A) which gives the conducting sphere a velocity v as shown by the vector (B). By the right-hand rule of physics, the angular velocity vector points in the negative z-direction. The angular velocity in the z-direction crossed with the radius r in the radial direction produces a velocity v in the clockwise θ-direction using cylindrical coordinates {r,θ,z}.

$v_\theta = w_z \times r_r = -\omega r$

- [0029]

Referring to FIG. 11, the negative radial dipole moment p_r crossed with the negative velocity v_θ of the sphere produces a positive magnetic moment μ_z in the z-direction.

$\mu_z = p_r \times v_\theta = (-p_r)(-v_\theta) = pv$

- [0030]

Referring to FIG. 12, the magnetic field B_z in the vertical z-direction is dotted with the magnetic moment μ_z in the z-direction to produce a force F_z in the vertical z-direction on each conducting sphere (FIG. 13).

$F_z = \nabla(\mu_z \cdot B_z) = \mu_z \nabla B_z$

- [0031]

The magnetic field that is produced by the solenoid actually curves away and around. Thus there is a gradient of the field in the z-direction.

- [0032]

The force can also be expressed in tensor notation. The magnetic B field in the vertical direction is part of an electromagnetic 4×4 matrix Faraday tensor F ▢
▢β = t ▢ ▢ ▢r ▢ ▢ ▢ θ ▢ ▢ ▢ z F β α = α = t 0 0 0 0 ▢ α = r 0 0 Bz 0 ▢ α = θ 0 - Bz 0 0 ▢ α = z 0 0 0 0 ▢
which shows that the magnetic field is located in slot F^r_θ of the Faraday tensor. In tensor notation the subscripts and superscripts have to match up on both sides of the equation. Matching subscripts and superscripts on the same side of the equation cancel. In this case, the electric dipole moment is in the radial direction p_r. The velocity can be represented as a time derivative of the θ-coordinate x^θ or $v_\theta = \partial x \theta \partial t$

- [0033]

Thus the force component in the z-direction becomes F z = p r ▢ v θ ▢ F θ , z r = (- p) ▢ (- v) ▢ ∂ B z ∂ z = pv ▢ ∂ B z ∂ z
where the angular and radial tensor components cancel and comma-z (, z) represents differentiation of the magnetic field in the z-direction.

- [0034]

The spacecraft design also has an inherent motion control system for moving in various directions. If the magnetic field of one solenoid arm is increased or decreased, the force on that sphere will be increased or decreased. Thus the spacecraft can turn in a particular direction.

A BRIEF DESCRIPTION OF THE DRAWINGS

- [0035]

FIG. 1. Electric dipole.

- [0036]

FIG. 2. Uniform electric field perturbed by electric dipole.

- [0037]

FIG. 3. Perspective view of spacecraft.

- [0038]

FIG. 4. Angle of solenoid tube.

- [0039]

FIG. 5. Electric field perturbed by conducting sphere.

- [0040]

FIG. 6. Electric dipole generated by conducting sphere.

- [0041]

FIG. 7. Orthogonal vector components of electric dipole.

- [0042]

FIG. 8. Magnetic field produced by tube arm solenoid.

- [0043]

FIG. 9. Orthogonal vector components of magnetic field.

- [0044]

FIG. 10. Angular velocity of hull.

- [0045]

FIG. 11. Magnetic moment produced by radial electric dipole and sphere velocity.

- [0046]

FIG. 12. Dot product of the magnetic moment with the magnetic field.

- [0047]

FIG. 13. Vertical lift force on all four conducting spheres.

- [0048]

FIG. 14. Perpective view of spacecraft interior.

DETAILED DESCRIPTION OF THE INVENTION

- [0049]

1. Referring to the cut-away view FIG. 14, the construction of the spacecraft is a thin-wall insulating thermoplastic having a dielectric constant in the range of 20 kilovolts per millimeter (A). An insulated electrode (B) runs from the cabin power supply and high-voltage transformer (C) to the tip of each tower (D). The four spheres (E) are silver plated to make them conducting. The tube solenoids (F) are driven by a direct current power supply (G).

- [0050]

2. The present model uses 3D computer design software and stereolithography fabrication techniques to create the thin-wall, low-weight, hollow structure of the hull. The computer model is sliced into many thin horizontal slices. A laser, mounted on an x-y table, draws out the slice on a table immersed in a bath of liquid polymer. Due to its sensitivity to the light, the liquid polymerizes. The table is then lowered a few thousandths of an inch more and the process is repeated. Thus making hollow spherical and conical shapes is extremely easy to do. Parts can be designed and stored in *.STL stereolithography files for transmission by Internet e-mail to the service bureau machine shop which sends the finished parts back the next day by express mail.

1. A spacecraft comprising:

a. a spherical cabin;

b. an electrostatic conical tower mounted on top of item (**1** *a*), supporting a vertically-mounted negatively-charged insulated electrode at the tip of the tower;

c. an electrostatic conical tower mounted on the bottom of item (**1** *a*), supporting a vertically-mounted positively-charged insulated electrode at the tip of the tower;

d. a vertical electric dipole created by items (**1** *b*) and (**1** *c*);

e. a high-voltage transformer to drive item (**1** *d*), mounted in item (**1** *a*);

f. four tubular arms, mounted at 90° around and extending at an angle from item (**1** *a*);

g. four solenoids, each of which is mounted axially inside item (**1** *f*);

h. a direct current power supply to drive item (**1** *g*);

i. four silver-plated conducting spheres, each of which is mounted on the end of item (**1** *f*);

2. an electrostatic lift system that:

a. produces a uniform spherical electric field by means of item (**1** *d*) which envelopes item (**1** *i*);

b. produces a perturbed electric field due to the presence of item (**1** *i*);

c. produces an electric dipole moment in the direction of item (**1** *b*) due to items (**2** *a*) and (**2** *b*);

d. produces a vertical magnetic moment due to the clockwise angular velocity of item (**1** *a*) combined with item (**2** *c*);

e. produces a vertical lift force on item (**1** *i*) due to item (**2** *d*) combined with the magnetic field gradient in the vertical direction produced by item (**1** *g*); and

f. creates a motion control system by varying the current to item (**1** *g*) in order to increase or decrease the effect of item (**2** *e*) on a particular item (**1** *i*).

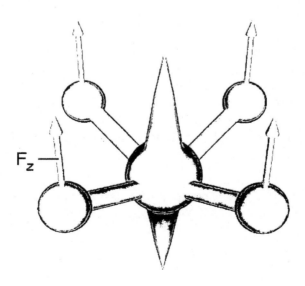

Figure 1

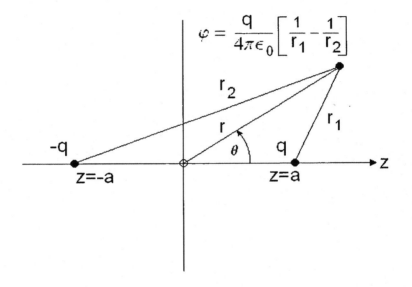

Figure 2

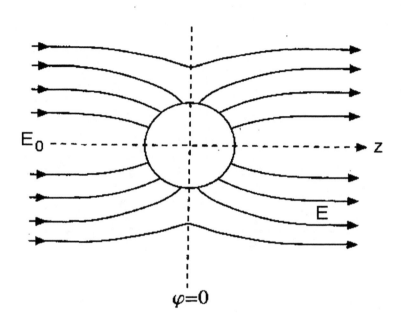

Figure 3

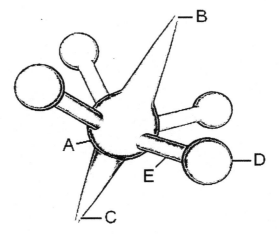

Figure 4

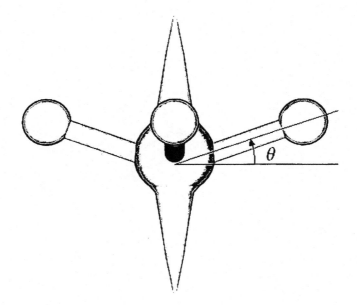

Figure 5

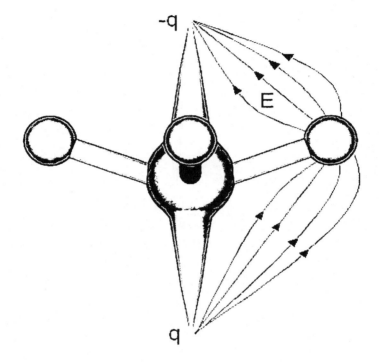

Figure 6

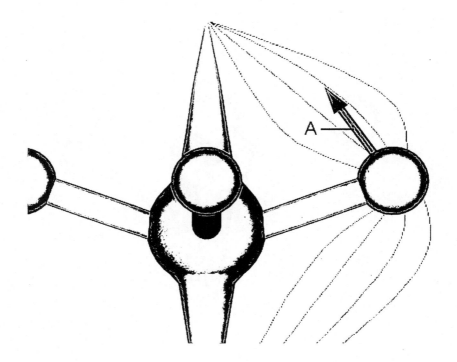

Figure 7

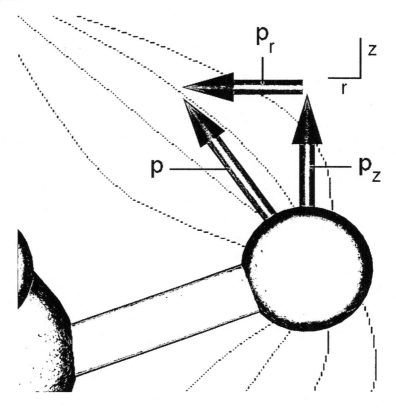

Figure 8

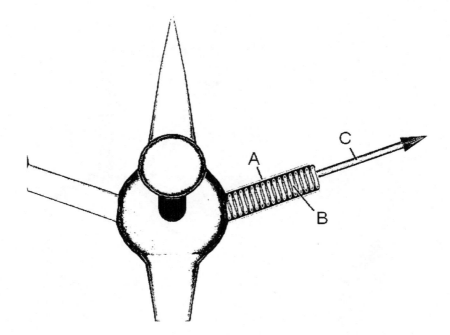

Figure 9

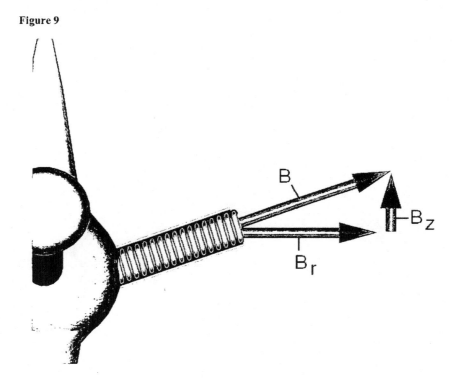

Figure 10

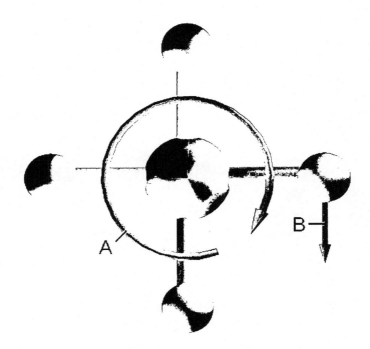

A

B

Figure 11

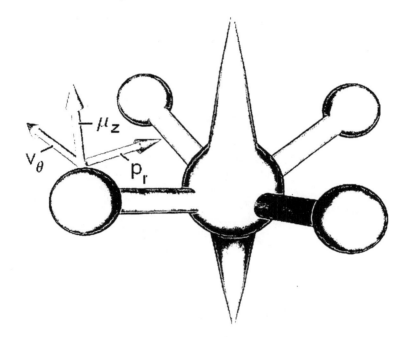

Figure 12

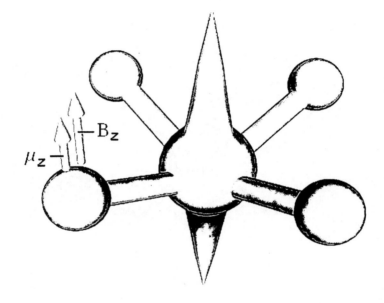

Figure 13

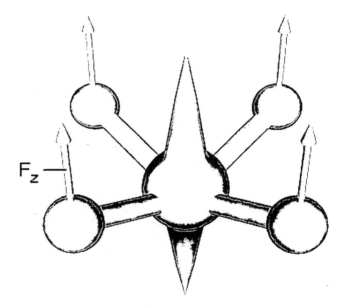

Figure 14

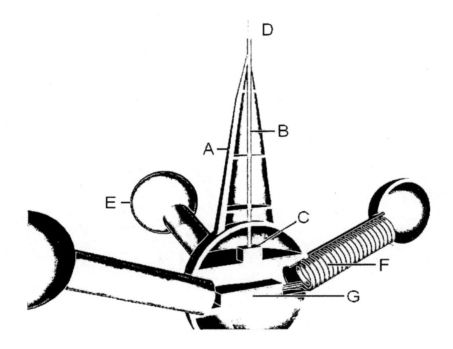

Chapter 4: Internet Accessible Mail Box System

Abstract

An Internet-embedded mail box system which transmits mail and package delivery information to a central server that can be accessed by wireless communication devices to determine if any mail is ready for pickup.

Classifications

G01V8/12 Detecting, e.g. by using light barriers using one transmitter and one receiver

US20060044139A1
United States

Inventor
   John St. Clair
   Michael Panzardi
Current Assignee
   Hecht Louis A

Worldwide applications
2004 US

Application US10/925,195 events

2004-08-24

Application filed by St Clair John Q, Michael Panzardi

2004-08-24

Priority to US10/925,195

2006-03-02

Publication of US20060044139A1

2015-12-29

Assigned to HECHT, LOUIS A

Status

Description

BRIEF SUMMARY OF THE INVENTION

- [0001]

  This invention is a commercial mail box system that transmits mail and package delivery information to an Internet server which can be accessed by wireless communication devices to know when mail is ready for pickup.

BACKGROUND OF THE INVENTION

- [0002]

  At the present moment, a person who rents a commercial mail box is forced every single time to have to personally visit the mail box rain, wind, storm, traffic jam or no parking in order to determine whether or not he or she has mail in the box. With the advent of wireless communication devices such as laptop computers, personal data assistants and Internet cellular telephones, a person can access an Internet website with these portable communication devices in order to obtain mail and package delivery information. As described in this invention, a mail box is fitted with an optical, electronic transmitting device which detects whether or not any letters or package delivery notices are in the box. If so, the device transmits this information to the Internet server which the user can access to determine whether or not it is necessary to travel all the way there for pickup. Thus the invention avoids unnecessary trips, summons for double parking, traffic congestion, and saves gasoline, time and money. By knowing that the package has been delivered before making a trip, usually in vain, personal stress is avoided. By knowing that the package the person was expecting has not arrived, the person can take that time to go to lunch or return home after a long and tedious workday.

SUMMARY OF THE INVENTION

- [0003]

  Referring to FIG. 1, the invention consists of a frame (A) containing mail boxes (B) attached to each other physically and electronically. Referring to FIG. 2, an individual box (A) contains two rectangular cutouts (B,C) centered in the bottom edge of the box. Referring to FIG. 3, an electronic circuit board (C) is attached by aluminum offsets to the bottom of the box (A). A light emitting diode (LED) (D), which forms part of the circuit board, sends a beam of light to the photocell (E) located directly across from (D) in the opposite slot in the box.

- [0004]

  Referring to FIG. 4, looking along the main axis of the mail box (A), containing bottom edge slots (B,C), a piece of correspondence (I) interrupts light beam (H) emitted from LED (F) so that photocell (G) receives no light. The resistance of the photocell increases which is detected by circuit board (D) and electronic board components (J).

- [0005]

Referring to FIG. 5, the detector circuit consists of a 5 volt supply with a voltage divider consisting of resistor (R1) connected to the supply, and photocell (R2) connected to ground (0V). Semiconductor chip (Maxim DS2401) detects the voltage on line (3). If the photocell receives light, the resistance decreases and line (3) detects zero voltage which is interpreted as a logic 0 level on the one wire data bus (2) corresponding to no mail. If the photocell receives no light corresponding to having a piece of correspondence in the mail box, the resistance increases and line (3) detects the supply voltage which is interpreted as a logic 1 level on the one wire data bus (2). All mail boxes are connected to the same one wire bus (2).

- [0006]

Referring to FIG. 6, all the mail box one-wire DS2401 switches (U1,U2) are connected in parallel to the Dallas TINI microcontroller through the One Wire Bus and the ground line (Gnd). Each switch has a unique identification number burned into its memory during manufacture. During manufacture of the mail boxes, the TINI microcontroller determines all the identification numbers and their corresponding mail box numbers using a binary tree search of the switches. These two corresponding numbers are stored in non-volatile memory in the TINI microcontroller together with the identification number of the TINI. At a predetermined time each day, the microcontroller polls each identification number on the one wire bus to determine the status of the data line (2) to see if there is any mail in a particular box. This information is coded in a record in memory as follows:

- [0007]

TINI microcontroller six digit identification number (XXXXXX)

- [0008]

Mail box four digit identification number (XXXX)

- [0009]

Date (MMDDYY)

- [0010]

Referring to FIG. 7, the abovementioned record for each mail box having mail is transmitted over an Ethernet cable from the TINI microcontroller to the modem (Hub). A telephone cable or wireless antenna/satellite connects the modem to the Internet. The microcontroller implements the Transmission Control Protocol/Internet Protocol (TCP/IP) networking Java communication software interface in order to connect to the Internet Service Provider (ISP). The TCP manages the assembling of a message or file into smaller packets that are transmitted over the Internet and received by a TCP server program that reassembles the packets into the original message. The IP handles the address part of each packet so that it gets to the right destination.

- [0011]

Referring to FIG. 8, the information in the abovementioned record is transmitted to the Mail Box Server through the ISP and server hub. The server implements a Java client listener software routine which detects these incoming transactions and stores them in a transaction file located on hard disk. The server is a large mail box database system consisting of the following files:

- [0012]

  (1) Client File

  - ○
    - (a) Name
    - (b) Mailing Address
    - (c) Billing Address
    - (d) Telephone Number
    - (e) Store Identification Number
    - (f) Mail Box Identification Number
    - (g) Password
    - (h) Phone Card Service (Yes or No)
    - (i) Client Identification Number

- [0022]

  (2) Mail Delivery Transaction File

  - ○
    - (a) Store Identification Number (TINI ID)
    - (b) Mail Box Identification Number
    - (c) Date

- [0026]

  (3) Package Delivery File

  - ○
    - (a) Store Identification Number (TINI ID)
    - (b) Mail Box Identification Number
    - (c) Mail Delivery Company (UPS, FEDEX, USPS, DHL)
    - (d) Envelope or Package Code (E, P)
    - (e) Tracking Number

- (f) Charges due

- (g) Date

- [0034]

  (4) Billing File

  - ○

    - (a) Client Identification Number

    - (b) Date of transaction (YYMMDD)

    - (c) Amount

    - (d) Description of transaction

    - (e) Transaction type (debit, credit)

    - (f) Check number

- [0041]

  The Mail Box server also processes user inquiries using a standard Internet program such as Microsoft's Internet Explorer with a website name such as www.mailbox.com. On a wireless laptop or desktop computer, the user enters his or her identification number and password. The server then looks up the client record and verifies the password. If the password is valid, the server uses the store and mailbox identification numbers found in the client file to access the mail and package delivery files whose information is formatted and displayed on the user's screen.

  A BRIEF DESCRIPTION OF THE DRAWINGS

- [0042]

  FIG. 1. Perspective view of mail boxes.

- [0043]

  FIG. 2. Perspective view of individual mail box.

- [0044]

  FIG. 3. Perspective view of mail box, circuit board, photocell and LED.

- [0045]

  FIG. 4. End view of mail box showing interruption of LED light beam.

- [0046]

  FIG. 5. Schematic drawing of circuit board.

- [0047]

  FIG. 6. Schematic drawing of TINI microcontroller and one-wire switches.

- [0048]

  FIG. 7. Schematic drawing of microcontroller connected to Internet.

- [0049]

  FIG. 8. Schematic drawing of Mail Box server connected to Internet.

- [0050]

  FIG. 9. Schematic drawing of dual port switch and LED.

- [0051]

  FIG. 10. Drawing of thin aluminum sheet for making mail box.

- [0052]

  FIG. 11. Perspective view of 1-wire bus connecting circuit boards.

- [0053]

  FIG. 12. Schematic drawing of package delivery system.

  DETAILED DESCRIPTION OF THE INVENTION

- [0054]

  1. Referring to FIG. 9, the Maxim DS2406 semiconductor is a dual port switch having one extra data port (PIOB, line **4**) compared to the single port DS2401. This extra port allows the TINI microcontroller to activate the light emitting diode (LED) only when the mail box is read. The microcontroller places a logic 0 on line **4** which grounds the LED. Current flows through resistor R3 and the forward biased LED such that light is emitted toward photocell (**R2**). The microcontroller then reads line **3** to determine if there is mail in the box. The LED is turned off by placing a logic 1 on line **4**. Because the LED is used for only a short time period each day, the life of the diode is increased substantially.

- [0055]

  2. Referring to FIG. 10, the mail box is cut with metal shearing equipment from a sheet of thin aluminum. A die and punch are used to cut out the rectangular holes (F) for the optics. Two holes (G) are threaded to attach the circuit board to the box using an aluminum offset and screw. The left side (A), bottom (B), right side (C), top (D) and riveting flange (E) are creased along four lines (H) to form the box. The box and attached circuit board slide into the mail box frame as a single unit.

- [0056]

3. Referring to FIG. 11, the circuit boards (A) are attached physically to their respective mail box using a spacer and screw (B). The aluminum spacer acts as a ground line to the electrically grounded mail box frame. An RJ-11 board connector on each circuit board routes the one-wire bus and +5V power supply line (D) sequentially through all the boards and to the microcontroller (E).

- [0057]

4. Referring to FIG. 12, a liquid crystal display/numeric keypad (LCD) and printer are connected to the TINI microcontroller to obtain package delivery information. The mail box attendant enters the box number, the shipping company such as UPS, FEDEX, USPS or DHL, the type of package such as envelope or package, date of arrival, charges due, and the tracking number. After completing the information, a delivery notice is printed out on the printer for insertion into the mail box. After adding the TINI identification and mail box number to the record, the microcontroller transmits the information over the Internet to the mail box server. This information is then stored in the aforementioned package delivery file.

- [0058]

5. As an additional service, the Mail Box server contains a voice response communications card which allows the server to call the user over a telephone modem and provide mail and package delivery information by electronic voice mail.

Claims

1. An Internet-embedded mail box system comprising:

a. a rectangular mail box frame;

b. item (a) containing a plurality of individual slide-in rectangular mail boxes made of folded thin aluminum sheet fitted with a swinging key-locked door on one side and open on the other for inserting pieces of correspondence into the boxes;

c. each mail box of item (b) containing two rectangular slots on opposite sides in the center bottom edges or other convenient location in order to provide an optical path from one slot to the other;

d. an electronic circuit board attached by aluminum spacers and screws to the bottom of each mail box of item (b);

e. an electronic circuit residing on item (d) comprising a light emitting diode (LED) on one side and a photocell on the other which provides a means of emitting and detecting a light beam across the mail box using the slots of item (c);

f. item (e) comprising a means of detecting a piece of correspondence in the mail box by interruption of the light beam;

g. item (e) comprising a +5 volt voltage divider consisting of a resistor and resistive photocell;

h. item (g) whose dividing voltage between the two components is measured by the first port on a two port switch, preferably a Maxim DS2406 chip, having a unique access identification number burned into its memory during manufacture, which transmits the logic level of one or zero, corresponding to the measured voltage level, along a 1-wire bus to a microcontroller connected by modem to the Internet;

i. item (h) logic level **1** corresponding to having mail, and logic level **0** to having no mail;

j. item (e) comprising a resistor from the power supply to the light emitting diode (LED) which in turn is connected to the second port of the dual port switch item (h) for the purpose of activating and deactivating the LED by means of the switch's unique identification number transmitted through the 1-wire bus by the microcontroller item (h);

k. item (h) microcontroller which transmits, using TCP/IP protocol and an Internet Service Provider (ISP), a record consisting of the microcontroller number (store number), mail box number and date to the mail box server which stores the information on hard disk in a transaction file;

l. item (h) microcontroller connected to liquid crystal display, numeric keypad and printer for the purpose of collecting and printing package delivery information which is transmitted by the microcontroller to the mail box server item (2) below;

m. a Java binary-tree search program used during manufacture to relate the mail box number with the unique identification number of the dual port switch of each mail box for storage in the microcontroller's non-volatile memory;

2. a Mail Box server comprising:

a. a multi-processor RAID computer system connected to the Internet by modem;

b. a database system running on item (a) consisting of the following files:

1. client master file,

2. mail delivery transaction file,

3; package delivery transaction file,

4. billing file;

c. a Java listener software program which listens for and processes transactions transmitted over the Internet by the microcontrollers;

d. a Java software program which processes client mail delivery inquiries through the Internet, using a client identification number and password, and displays the information on the client's screen in order for the client to know if any mail is ready for pickup.

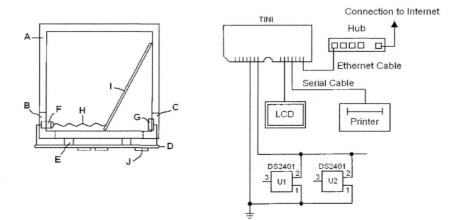

Figure 1

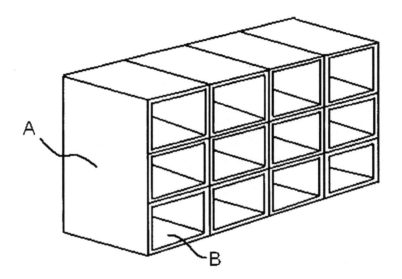

Figure 2

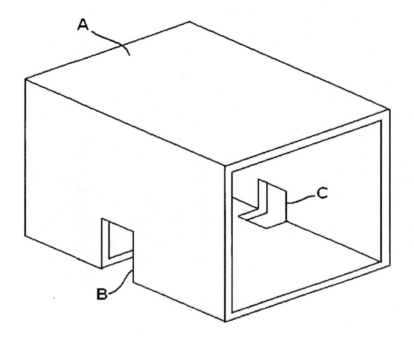

Figure 3

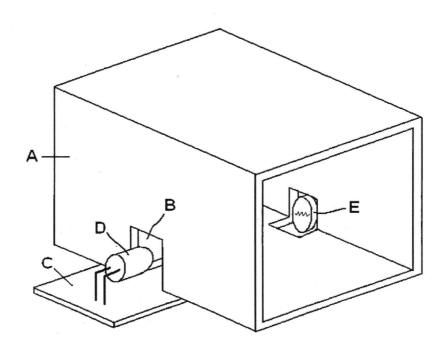

Figure 4

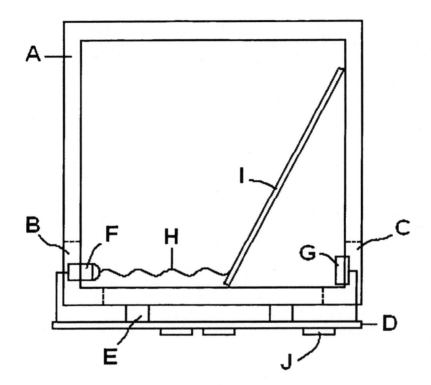

Figure 5

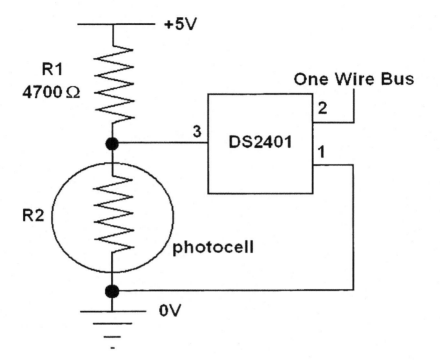

Figure 6

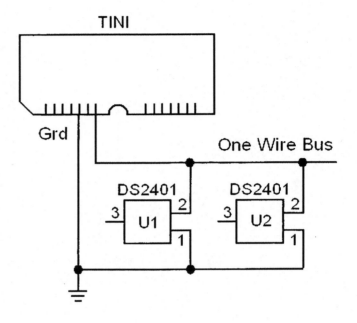

Figure 7

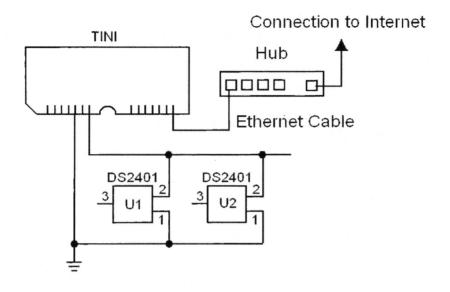

Figure 8

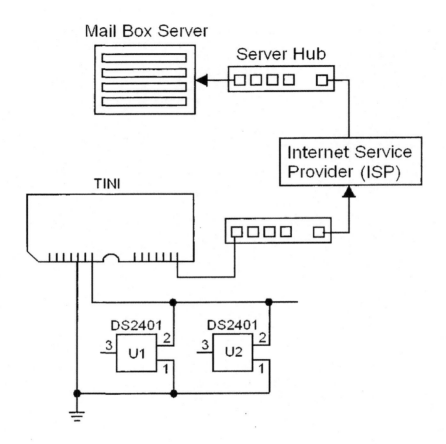

Figure 9

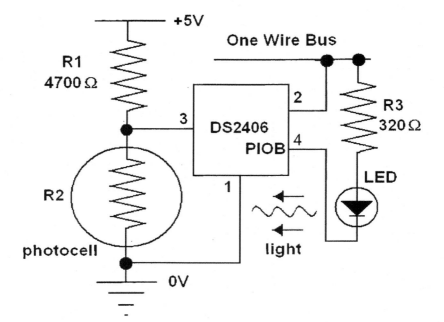

Figure 10

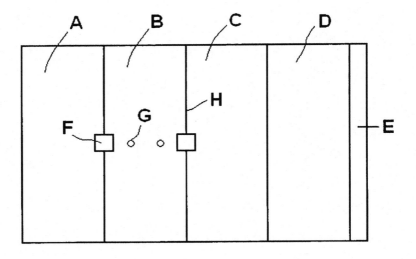

Figure 11

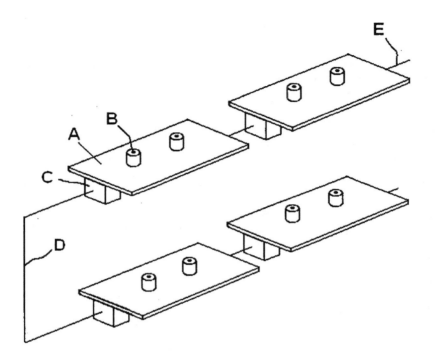

Figure 12

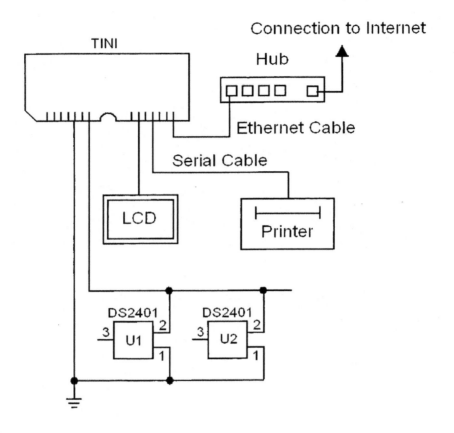

Chapter 5: Full Body Teleportation System

Abstract

A pulsed gravitational wave wormhole generator system that teleports a human being through

Classifications

B64G1/409 Unconventional spacecraft propulsion systems

US20060071122A1
United States

Inventor
  John St. Clair
Current Assignee
  HECHT ROJAS LUIS A

Worldwide applications
2004 US

Application US10/953,212 events

2004-09-29

Application filed by St Clair John Q

2004-09-29

Priority to US10/953,212

2006-04-06

Publication of US20060071122A1

2015-07-07

Assigned to HECHT ROJAS, LUIS A

Status

Abandoned

Description

BRIEF SUMMARY OF THE INVENTION

- [0001]

This invention is a system that teleports a human being through hyperspace from one location to another using a pulsed gravitational wave traveling through hyperspace.

BACKGROUND OF THE INVENTION

- [0002]

The basis for this invention is an event, referring to FIG. 1, occurring on May 2, 2004, in which the inventor ("he") personally experienced a full-body teleportation while walking to the bus stop (A) along a road (B) that runs perpendicular to the nearby commercial airport runways where planes are landing. There is a wide iron grating (D) for water drainage that crosses the road at the center of the bus stop. The grating width is such that one has to make a concerted effort to jump across it in order to get from one side to the other. Approximately 50 meters from the iron grating, he (E) felt a vertical wave (F), similar to a flag waving in the breeze, traveling down the street toward the bus stop. The wave velocity was about 1 meter per second, which was slightly faster than his walking speed. In the next instance, he (G) found himself down the street near the corner of the next block. Realizing that he had passed the bus stop, he turned around to see the iron grating approximately 50 meters up the street in back of him. Because there was no recollection of having jumped across the iron grating nor of having passed the bus stop's yellow marker line, he realized that he had been teleported a distance of 100 meters while moving along with the traveling wave. It was obvious that the wave was pulsed because the front edge overtook the inventor, moved with him momentarily, and then the back edge of wave left him as it moved on down the street. While contemplating this sequence of events, he then looked up and saw in a span of a few seconds a twin-turboprop airplane (C) in the distance crossing above the road while making a shallow descent in order to land at the airport.

- [0003]

It took a number of days in order to understand this sequence of events. The explanation involves knowledge of a wide range of subjects such as gravitation physics, hyperspace physics, wormhole electromagnetic theory and experimentation, quantum physics, and the nature of the human energy field.

- [0004]

It is obvious from the above scenario that the airplane momentarily crossing perpendicular to the road generates the aforementioned pulse. Because the airplane has an engine on each wing, there are two propellers which conceivably are rotating out-of-phase with each other. That is, the blade of one propeller could be pointing up and the equivalent blade on the other engine could be pointing in a slightly different direction. Notice that the tip of the blade traces out a helix as the plane is landing.

- [0005]

In gravitation physics, referring to FIG. 2, it is known that two masses of mass m1 and m2 (A,B) attached by lever arms slightly offset by an angle δθ along the radial direction to the rotating shaft (C), will produce a gravitational wave (D) traveling perpendicular to the shaft. The mass and wave are referred to as the source and receptor respectively. Referring to a side view looking along the shaft FIG. 3, the product of the mass m times the angular acceleration a is a constant such that m1 *a* 1 is equal to m2 *a* 2. The distance between the masses is length L, which makes an angle θ with the horizontal axis. The difference in time of travel to the receptor gives rise to a difference in phase δθ equal to the angular velocity ω of the rotating shaft times the length L times the cosine of the angle θ

$$\delta\theta = \omega L \cos(\theta)$$

- [0006]

At the receptor, the amplitude of the wave is equal to the mass times the acceleration times the phase difference divided by the radius r to the receptor A = m 1 ⊠ a 1 r ⊠ δθ ≈ (m ⊠ ⊠ ⊠ ω ⊠ ⊠ ⊠ L ⊠ ⊠ ⊠ sin ⊞ (θ) r) ⊠ (ω ⊠ ⊠ ⊠ L ⊠ ⊠ ⊠ cos ⊞ (θ)) = m ⊠ ⊠ ⊠ ω 2 ⊠ L 2 ⊠ sin ⊞ (2 ⊠ θ) r
Even though the turboprop airplane engines have a high rotational speed and a large separation distance between masses, the gravitational wave which is produced is small and not noticed. The problem is that the gravitational constant G in this dimension has such a small value equal to the speed of light c squared divided by the linear mass Ω of the universe G = c 2 Ω = (299792458 ⊠ ⊠ ⊠ m ⊠ / ⊠ s) 2 1.346812891 · 10 27 ⊠ ⊠ ⊠ kg ⊠ / ⊠ m = 6.673200002 · 10 - 11 ⊠ ⊠ ⊠ m 3 kg ⊠ ⊠ ⊠ s 2

- [0007]

On the other hand, a gravitational wave traveling in hyperspace would be magnified enormously due to the face that the linear mass is so small. The magnitude of the gravitational constant in hyperspace can be estimated in the following manner. At the beginning of the 20th century, a man's parents were dying of tuberculosis. With their permission, he placed them and their beds on weighing scales. When each one passed away, each scale registered a drop in mass equal to 0.071 kilograms. This is the mass of the hyperspace energy being which resides in the physical body. Because hyperspace is co-dimensional with our dimension, the energy being interpenetrates the body and controls its movement.

- [0008]

Referring to FIG. 4, a human being has seven vortices (A through G) which are aligned along the centerline of the body. Each vortex is actually a co-gravitational field K which causes a pendulum placed in the field to spin in circles. For this reason, the K field has units of inverse seconds similar to an angular velocity. The vortex transports energy from our dimension to the energy being located in hyperspace. The gravitational field g and the co-gravitational field K are equivalent gravitationally to the electric E field and the magnetic B field found in electromagnetism. The equivalent gravitational solution to an electromagnetic problem can be obtained by substituting the following gravitational constants for the electromagnetic constants Electromagnetic Gravitational q (charge) m (mass) ρ (volume charge density) ρ (volume mass density) σ (surface charge density) σ (surface mass density) λ (line charge density) λ (line mass density) J (convection current density) J (mass current density) E (electric field) g (gravitational field) B (magnetic field) K (co-gravitational field) ε_0 (permittivity of space) $-\frac{1}{4}\pi G$ μ_0 (permeability of space) $-4\pi G/c^2$ $-\frac{1}{4}\pi\varepsilon_0$ or $-\mu_0 c^2/4\pi$ G (gravitational constant)

- [0009]

Referring to FIG. 5, each vortex is connected through the pineal gland by light cords to a separate hyperspace quantum well having its own frequency and dimension. The reason for this separation is that the conical spiritual eye, attached to vortex B, has to have its own energy structure which is different from vortex (A) which is connected to the quantum energy field in which the mental processes are developed. Due to the high speed of light in our dimension, the quantum wells are the size appropriate to molecules and atoms. In hyperspace, where the speed of light is one meter per second, the quantum wells are huge and can be manipulated. This manipulation has shown that the quantum wells are in the shape of a cube about a meter on a side. This makes the whole structure about seven meters tall for a volume of seven cubic meters. Thus the mass density ρ of hyperspace is about ρ h = .071 ▨ ▨ ▨ kg 7 ▨ ▨ ▨ m 3 = .01 ▨ ▨ ▨ kg m 3
which per unit area is the same value. Therefore the hyperspace gravitational constant is equal to G h = c 2 Ω h = (1 ▨ ▨ ▨ m s) 2 .01 ▨ ▨
▨ kg ▨ / ▨ m = 100 ▨ ▨ ▨ m 3 kg ▨ ▨ ▨ s 2
The enormous magnification of the gravitational constant is therefore of the order of G h G = 100 6.6732 · 10 - 11 ≈ 1.5 · 10 12
The question is how does this amplified gravitational wave created by the rotating propellers and turbines get into hyperspace from our dimension?

- [0010]

  The answer comes from experiments done using the ancient Chinese form of breathing known as Chi Kung. Using this breathing technique, we have been able to levitate the human body over six feet in the air. The internal temperature of the stomach is around 200 degrees Fahrenheit. By simultaneously squeezing the diaphragm to bring hot air up through the lungs, and breathing through the nose to bring cold air down, rotating vortices are generated in the lung passages when these two air masses meet and twist around each other as depicted in the famous Yin-Yang diagram. Because the lung has variable diameter passages from the large diameter at the throat to the final small air sacs, there is a spectrum of rotating frequencies.

- [0011]

  From quantum physics it is known that if there is a temperature fluctuation occurring among a group of harmonic oscillators in the environment, then Planck's reduced constant

  $\hbar$

  is increased by the cotangent of the constant times the frequency ω of the oscillator divided by twice Boltzmann's constant k times the temperature T $\hbar = \hbar$ ▨ ▨
  ▨ coth ▨ ($\hbar\omega$ n 2 ▨ kT)

- [0012]

The effect of increasing Planck's constant, referring to FIG. 6, can be seen in the tetrahedron diagram. This diagram, of which there are now over 4000, plots the natural logarithm of mass on the vertical axis versus the natural logarithm of wavelength on the horizontal axis. In terms of mathematics, it is a subspace logarithmic manifold which projects geometrically the physics constants into our **4D** spacetime dimension. That is, it is the geometry of the tetrahedron circumscribed by the sphere that determines the mass of the proton and electron. The mass of the electron times its wavelength is equal to the mass of the proton times its wavelength which in turn is equal to Planck's constant h divided by the speed of light c m e ▨ λ e = m p ▨ λ p = h c

Taking the natural logarithm of the above equation shows that the mass plus the wavelength is equal to what is termed the base

constant ln ▨ (m e) + ln ▨ (λ e) = ln ▨ (h c) = - 95.91546344

which is represented in FIG. 6 by the 45 degree line (A) from point (a) on the horizontal axis to the vertical axis at point (b). The electron is located at point (c) which is the intersection of the electron wavelength (B) with line (A). The electron wavelength (B) reflects off the sphere (D) at points (d) and (e) and returns along line (C) as the electron mass. As shown in tetrahedron diagram tet0565, stored in the Library of Congress, the clockwise path of the electron transitions into the counter-clockwise path of the proton showing that the electron and proton are one and the same particle. Because the electron and proton travel in opposite directions along the path, they have the same charge but of opposite sign.

- [0013]

Our dimension is represented by Planck box (E) which is bounded by the Planck mass and the Planck wavelength. The Planck mass is equal to the linear mass of the universe times the Planck length which is the bottom limit of our dimension. The Planck wavelength is 2π times the Planck length. Notice that the electron is located within the Planck box.

- [0014]

Referring to FIG. 7, if there is an increase in Planck's constant due to the temperature fluctuations among the harmonic oscillators, the 45 degree base line (A) moves to the left on the tetrahedron diagram as shown by line (F). Because of the increase in the base constant, there is a corresponding increase in the electron mass and wavelength. The electron moves from point (c) to point (f) which places it at the edge of the Planck box (F) which is the boundary between space and hyperspace. At point (f), the electron is essentially no longer in our dimension.

- [0015]

Referring to FIG. 8, imagine a box (A) filled with nine electron oscillators (B). If Planck's constant is increased near the three oscillators in the middle, these electrons will leave this dimension. This leaves six oscillators as shown in the box (C). However, box (C) is the equivalent of box (D) in which there are still nine positive mass oscillators together with 3 negative mass oscillators. Thus there is an accumulation of negative energy (−ρ) when information is lost from the environment to another dimension.

- [0016]

Dr. Kip Thorne, who co-authored the book *Gravitation* with Dr. Archibald Wheeler of Princeton University, has shown in a General Relativity spacetime curvature calculation that negative energy is required to open and stabilize the throat of a wormhole between space and hyperspace. The accumulation of negative energy in the aforementioned example generates wormholes between into hyperspace. Hyperspace has a low energy density because of the reduced speed of light in that dimension. Ordinarily, energy would not flow from hyperspace to space because space has a higher potential than the potential of hyperspace. This, of course, is the reason that the body vortices can flow energy into the energy field of the human being who is located in hyperspace. By creating negative energy, the potential becomes reversed such that low density hyperspace energy flows into our dimension as seen by the positive head

$PE = \rho_{hyperspace} - (-\rho_{space}) = +2\rho$

The low-density energy fills the body which allows a human being to float upwards like a helium balloon as verified by Chi Kung breathing as well as spinning on a motorized platform known as the Chakra Vortex Accelerator. The latter device resulted in the first mechanical means to produce anti-gravity.

- [0017]

The process of creating spinning thermal fluctuations is the same as found in the hot air vortices created by the jet airplanes landing at the airport near the road where the full-body teleportation occurred. Large vortices are created over the wing of the airplane at the same time that the turbine engines are spinning hot vortices into relatively cold air. These conditions produce wormholes between space and hyperspace. It takes a twin turboprop airplane landing behind the jet to generate the gravitational wave in the region where the wormholes have formed. The gravitational wave then traverses the wormholes into hyperspace, becoming highly amplified due to the change in linear mass and speed of light. Because the propeller blades are co-linear with the road, the gravitational wave travels in the direction along the road where it was encountered by the inventor.

- [0018]

From experiments with cavitating bubbles (see patent application Cavitating Oil Hyperspace Energy Generator), it was found that it is possible to produce a wormhole if the surfaces of the bubble collapse asymmetrically. A symmetric collapse of a spherical bubble produces enormous spacetime curvature distortions. An asymmetric collapse, using a magnetic field to distort the collapse, produces, in addition to the same severe spacetime distortions, negative energy as the bubble collapses. Due to some General Relativity considerations, the wormhole that is created starts rotating in a manner similar to the beacon light produced by a lighthouse.

- [0019]

Referring to FIG. 9, due to the forward helical motion (A) of the propellers (B) as the airplane crosses the road, the pulsed gravitational wave (C) is skewed backward at an angle (D). Due to the wormholes created by the presence of thermal vorticity fluctuations generated by the wing and turbines of the airplane, this skewed wave moves into hyperspace where it is highly magnified and detected by the inventor.

- [0020]

Referring to top view FIG. 10, the gravitational wave (A) causes a skewed compression and expansion of the hyperspace quantum wells (B) which constitute the human energy being. Due to this asymmetric distortion in the xy-plane, the quantum wells take the physical body out of dimension as long as the wave pulse is traveling with the human energy field. Once the back edge of the gravitational wave moves on past the quantum wells, the body is then brought back into dimension.

SUMMARY OF THE INVENTION

- [0021]

It is the object of this invention to teleport a human being from one location to another by creating a pulsed gravitational wave traveling through hyperspace that asymmetrically compresses and expands the quantum wells of the human energy being. This spacetime curvature distortion of the hyperspace quantum wells pulls the physical body out of dimension such that the human being is teleported along with the wave. As the pulsed wave moves on past the quantum wells, the human is brought back into dimension at some distant location. The invention requires (1) a device that will generate a wormhole between space and hyperspace, and (2) a device that will generate a gravitational wave which can be inserted through the wormhole.

- [0022]

Referring to FIG. 11, a magnetic vortex wormhole generator has already been developed which generates a wormhole between space and hyperspace as described in a previous patent application entitled Magnetic Vortex Wormhole Generator. Using this generator, it was found that smoke blown through one side of the coil does not appear on the other side of cylindrical coil. The smoke flows through the wormhole and appears in a hyperspace co-dimension. It was this experiment that resulted in making first contact with the androids of the Grey aliens who told me, in a remote viewing session, that "We saw you blowing smoke into hyperspace."

- [0023]

The wormhole generator consists of two concentric cylindrical coils (A,B), one of larger radius than the other, made of thin transformer iron laminate wrapped in opposite directions with one continuous wire driven by a sinusoidal current. The solenoidal coil generates a magnetic field through the laminate. Because the electrical current flows in opposite directions at different radii through the two windings, bucking electric fields (C) are created along the centerline of the generator. These radially-offset magnetic fields and bucking electric fields, as shown by a calculation using Einstein's General Theory of Relativity, generate both an enormous spiking spacetime curvature and negative energy at small radius along the centerline where the wormhole is formed. The gravitational wave generator is then coupled to this wormhole generator.

- [0024]

Referring to FIG. 12, it is known from gravitation physics that injecting an electromagnetic wave (A) into a hollow toroidal waveguide (B) produces a hyperbolic spacetime curvature stress (C) in the plane of the waveguide. The tips of the arrows indicate compression and the tail of the arrows indicate expansion or stretching of spacetime. The reason for this spacetime curvature is because the waveguide forces the electromagnetic wave to curve around and travel in a circle. Spacetime has to compensate for this toroidal-generated stress by creating hyperbolic lines of stress in the inner plane of the toroid so that the overall spacetime curvature is zero. For a greater gravitational effect, three toroidal waveguides, phased 120 degrees apart, are used to seal off the curvature.

- [0025]

Referring to FIG. 13, the three toroids create a rotating, twisting, vertical propagating gravitational field (A) through the centerline of the toroids provided that the period of the electromagnetic wave is twice the period of the gravitational wave. This phase relationship is adjusted by selecting the correct radius for the frequency of the monochromatic wave.

- [0026]

In order to effectively use this gravitational wave, referring to FIG. 14, three phased toroidal waveguides (A,B) are mounted at the top of each of two identical square granite obelisks (C,D). The two obelisks are offset by a short distance between them. As the vertical gravitational wave rotates around along the vertical axis inside the obelisk, the edges of the square obelisks are compressed and expanded such as to create two cylindrical asymmetric gravitational waves traveling radially outward.

- [0027]

Referring to FIG. 15, these waves meet to form a plane gravitational wave (A) which travels down the centerline between the two obelisks.

- [0028]

Referring to FIG. 16, the full body teleportation system consists of the twin granite obelisks (A,B) on which are mounted near the top of each the toroidal waveguides (C,D) which produce the pulsed gravitational waves (E,F) that run the length of the obelisks. Because the gravitational wave is rotating inside the obelisk, the granite stone undergoes a very small asymmetrical compression and expansion. A cylindrical gravitational wave propagates out from each obelisk such that along the centerline between the two there is generated a plane gravitational wave. This wave enters the wormhole (H) created by the magnetic vortex generator which is located a short distance from and parallel to the obelisks. The wave is amplified by a factor of almost 10^{13} when it enters the hyperspace co-dimension.

A BRIEF DESCRIPTION OF THE DRAWINGS

- [0029]

FIG. 1. Perspective view of site where full-body teleportation occurred.

- [0030]

FIG. 2. Perspective view of gravitational wave generator.

- [0031]

  FIG. 3. Planar view of gravitational wave generator.

- [0032]

  FIG. 4. Perspective view of seven vortices of human energy being.

- [0033]

  FIG. 5. Perspective view of seven large quantum wells of human energy being.

- [0034]

  FIG. 6. Tetrahedron diagram showing Planck's constant and electron.

- [0035]

  FIG. 7. Tetrahedron diagram showing electron moving out of dimension.

- [0036]

  FIG. 8. Perspective view showing production of negative energy.

- [0037]

  FIG. 9. Perspective view of skewed gravitational wave produced by propellers.

- [0038]

  FIG. 10. Planar view of skewed quantum wells deformed by gravitational wave.

- [0039]

  FIG. 11. Perspective view of magnetic vortex wormhole generator.

- [0040]

  FIG. 12. Perspective view of hyperbolic lines of stress generated by toroidal waveguide.

- [0041]

  FIG. 13. Perspective view of rotating, twisting, propagating gravitational wave generated by toroidal waveguides.

- [0042]

  FIG. 14. Perspective view of toroidal waveguides attached to obelisks.

- [0043]

  FIG. 15. Perspective view of gravitational wave generated by obelisks.

- [0044]

  FIG. 16. Perspective view of magnetic vortex wormhole generator and obelisk gravitational wave generator.

  DETAILED DESCRIPTION OF THE INVENTION

-

  - 1. The obelisks are quarried out of granite stone and cut with a large-diameter diamond saw that is used in highway construction. The beveled piece at the top is cut separately and cemented in place. A tapered aluminum bracket holds the toroids in place.

  - 2. The electronics for the magnetic vortex generator are similar to that used in the patent application Magnetic Vortex Wormhole Generator.

  - 3. The electronics for the toroidal waveguides is the familiar stub and coaxial cable driven by an amplifier and pulsed variable-frequency generator.

Claims

1. A full body teleportation system consisting of:

generating a pulsed gravitational wave which propagates through a magnetic vortex wormhole generator; and generating a wormhole with the magnetic vortex generator whereby the pulsed gravitational wave traverses through the wormhole and enters into hyperspace where the wave is enormously magnified due to the lower speed of light in that dimension.

2. The method of claim 1, wherein the step of generating the pulsed gravitational wave comprises:
using two granite stone obelisks;
mounting monochromatic-wave toroidal waveguides on top of each obelisk to create a rotating, twisting, propagating gravitational wave through the vertical axis of each obelisk; and
creating a cylindrical compression and expansion in each obelisk to produce a plane gravitational wave traveling down the centerline between the two obelisks.

3. The method of claim 1, wherein the step of generating a wormhole into hyperspace comprises:
using two concentric cylindrical solenoidal coils of different radii connected by a single wire wrapped in opposite directions on thin iron transformer laminate;
generating bucking electric fields down the centerline of the vortex generator which creates a spacetime curvature distortion with negative energy in accordance with Einstein's General Theory of Relativity.

4. A teleportation system comprising:
generating a gravitational wave traveling through hyperspace which interacts with the human energy being; and

pulling the human energy being and physical body out of dimension when interacting with the pulsed gravitational wave such that the person is teleported from one location to another through hyperspace and back again into our 4D spacetime dimension.

Figure 1

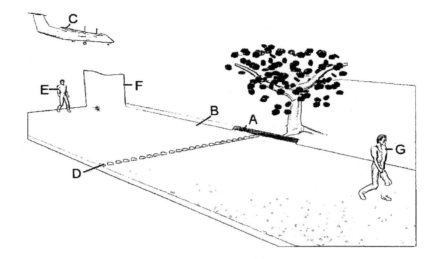

Figure 2

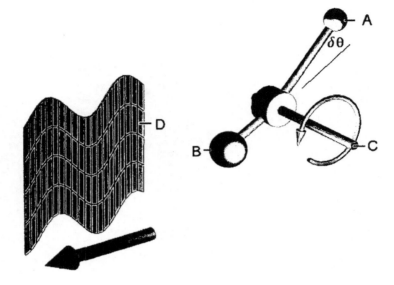

Figure 3

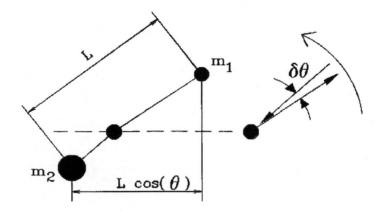

Figure 4

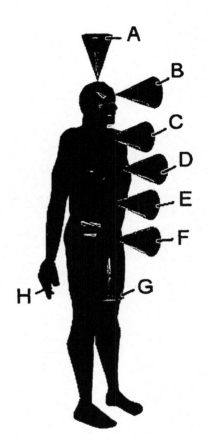

Figure 5

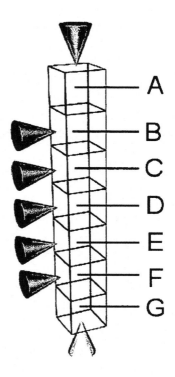

Figure 6

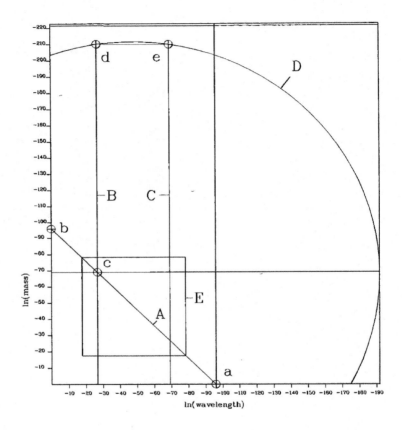

Figure 7

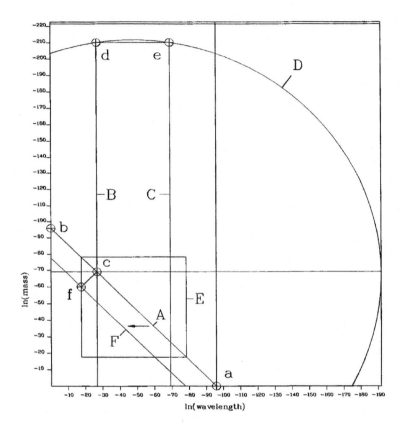

Figure 8

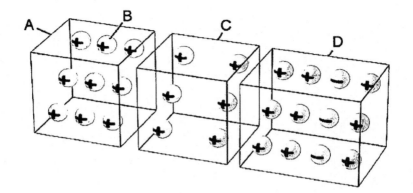

Figure 9

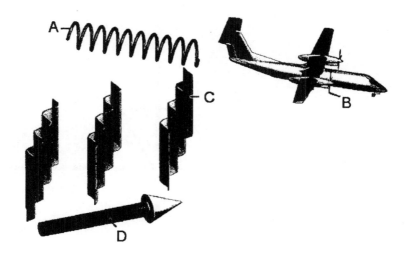

Figure 10

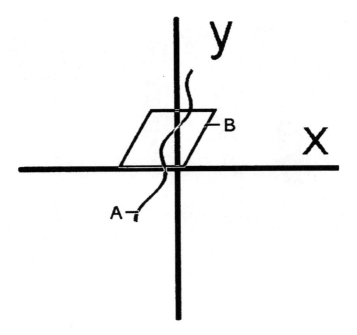

Figure 11

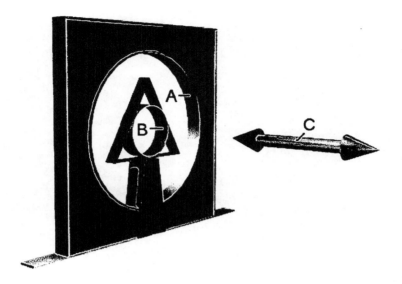

Figure 12

Figure 13

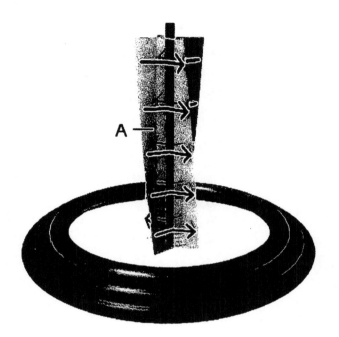

A

Figure 14

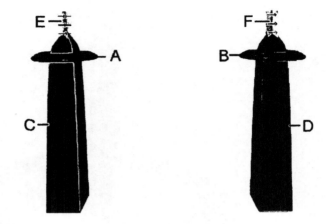

Figure 15

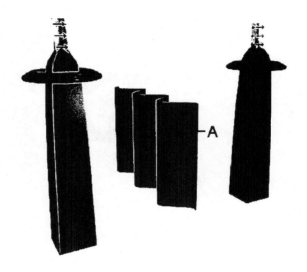

Figure 16

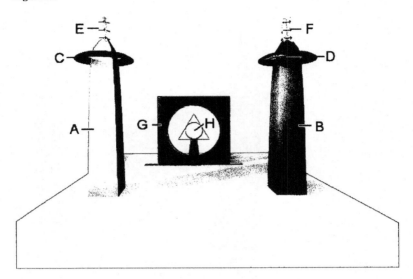

Chapter 6: Remote Viewing Amplifier

Abstract

An apparatus which enhances the ability of a person to perform remote viewing by connecting the human spiritual eye to the tetrahedral geometry of subspace.

Classifications

G02B27/02 Viewing or reading apparatus

US20060072226A1

United States

Inventor

John St. Clair

Worldwide applications
2004 US

Application US10/957,391 events

2004-10-02

Application filed by St Clair John Q

2004-10-02

Priority to US10/957,391

2006-04-06

Publication of US20060072226A1

Status

Abandoned

Description

BRIEF SUMMARY OF THE INVENTION

- [0001]

This invention enhances the ability of a person to perform remote viewing by connecting the spiritual eye to the tetrahedral geometry of subspace.

BACKGROUND OF THE INVENTION

- [0002]

Remote viewing is the projection of spiritual modules of the human energy field to distant locations in order to see, communicate and interact with other entities who live in subspace, space and hyperspace co-dimensions of the universe.

- [0003]

One of my first remote viewings was made at night to a distance of 10,000 miles on the sunlit side of the earth. My spiritual eye and body projected together while my mental facilities remained in my physical body. I found myself looking down on a palm tree from a height of about one hundred feet. The palm tree had several coconuts in it as seen in FIG. 1. I then gave the command to lower myself to the ground. At that moment I went sailing down past the coconuts, barely missing the tree! Finding myself on a pathway through the tropical forest, I then came to an extremely long wooden bridge which crossed over a river gorge. On the other side of the bridge I could see three soldiers running toward me as shown in FIG. 2. The two soldiers in front were carrying rifles and wearing light blue berets. The man running behind them was wearing an officer's cap with a red band. My first reaction was that I was going to be shot. I edged over on the right side of the wooden railing. They ran right past without seeing me. I then asked to see the building that these soldiers were guarding. Everything went dark, and then I found my spiritual eye peeking out of the floor of a computer room as seen in FIG. 3. There was one man using a computer on the opposite side of the room near an open door. He got up from his chair and came over to sit in front of a second computer located a few feet from where I was located. From the glare of the computer monitor, I could clearly see his face. Everything went dark as my spiritual eye and body projected back to my physical body.

- [0004]

Another time my spiritual eye, spiritual body and mind were standing outside the closed front door of my condominium. Upon patting my legs with my hands, I couldn't find the keys in my trousers. When I looked down, I realized I wasn't in my physical body. I then shot through two solid walls of concrete and returned to my awakening body.

- [0005]

What these two examples show is that the human spiritual energy system is modular. The reason it is modular is because there are seven hyperspace co-dimensions, each vibrating at a slightly different frequency, which receive energy from space through seven vortices located along the physical body. Over a lifetime, these vortices build up the human aura. Because all energy systems have to be grounded, the remaining six modules are the legs, body, arms/hands, voice, eye and mind. When the entire group is out-of-body as a single entity, then the soul energy powers the body in a manner similar to a battery. The soul looks like a two-inch diameter orange ball of plasma. If the soul is removed from the body, then the body becomes paralyzed except for a small movement of the eyelids. Upon death, all these separate modules are assembled into a single energy being. A light cord from the soul provides the battery power and information required to join these modules together. A second light cord from the pituitary gland transfers the modules into the energy being for assembly. This energy information transfer is the reason that people in a near-death situation say they saw their entire life flash before them.

- [0006]

Moving to a larger picture of things that are happening in the galaxy, I was able to make contact with the Pleiadian Federation which is located about 400 light years from earth. The Federation is a group of over one hundred intelligent beings that were brought to the Pleiades from around the galaxy. One member of the Federation calls itself the Intelligent Insect Beings. They are the ones who fly the black triangles over Belgium and France for the purpose of evacuating human beings back to the Pleiades for relocation on a planet called Earth II. The reason for this evacuation was that it was not known if it would be possible to win the battle of Revelations, which would take place about two years later here on earth.

- [0007]

As could be expected, the humans were angry and hysterical at being abducted. So the Intelligent Insect Beings asked me if I could calm them down. I was in telepathic communication with them, and they were in telepathic communication with their computer system, which meant that I could have my thoughts displayed to the humans on the computer monitor. It turned out that one woman was from Central America and only spoke Spanish. The Intelligent Insect Beings didn't speak Spanish so they thought there would be no way I could communicate with her. So I told them that I would spell the Spanish words letter-by-letter and she could then read my message. Since most Spanish people are Catholics, I thought a religious message would be of importance to her. I spelled out the phrase, "Que Dios te bendiga." which means, "May God bless you." I also asked the two beings to put their hands together in a form of prayer, and the woman followed suit. At that moment the Intelligent Insect Beings were reading her mind to understand her emotional state. They said, "She is crying tears . . . " After a slight but tense pause, they continued, "of joy!" They said the woman had a big smile on her face and was successfully transitioned into her new life.

- [0008]

One year later, the Blond Aliens of the hundred member council of the Pleiadian Federation remote viewed me so that I could celebrate with them the success of their mission to earth. The Blond Aliens, which is not their real name, fly the Beamship spacecraft. This hyperspace vehicle can teleport itself to any location in the galaxy. It dematerializes into the black void for an instance and then re-materialize anywhere in the galaxy. This ship can be seen in Billy Meier's video from Switzerland along with a picture of their envoy, Semjase.

- [0009]

  Just recently I had a demonstration of the Beamship's ability. I heard an emergency distress call by a commander of a spacecraft who said that they had a fire onboard their spacecraft. I immediately remote viewed a Federation maintenance mothership, they got the frequency and location of the ship, and dispatched the Beamship in time to rescue the commander. The burned-out cables, which had caused the fire, were repaired by the Federation and the commander arrived home safely on his planet in his own spacecraft.

- [0010]

  The Federation then received a message from the commander saying that he wanted his planet to join the Federation now that there was this type of communication available. So the Federation visited his planet for the signing ceremony, and I was invited to attend the proceedings by remote view. Word got around about this, and within three weeks another 20 planets joined the Federation for the same reason.

- [0011]

  At one ceremony, which was attended by Admiral Third Class of the Pleiadian Defense Department, His Highness of the planet was signing the document of incorporation. At that moment, I could see him signing, so I exclaimed, "He is signing with his left hand." The Admiral was almost apoplectic at hearing this. After giving her the signed document, His Highness held up his hand and asked me how many fingers he was holding up. He had a fist so I said none. Then he put his index finger out and I said one. He then made a fist again and I said none. Then he held out all fingers and I said five. The Admiral said that he had a big smile on his face as he went to announce the agreement because he knew, even though his planet was located **90,000** light years away on the other side of the galaxy, he could instantly communicate any problems to the Federation. So this is the importance of developing inventions that can enhance our remote viewing ability because one day it will mean that we can become a vital part of the Pleiadian Federation.

  SUMMARY OF THE INVENTION

- [0012]

  Referring to FIG. 4, the spiritual eye of the human energy system is located at the pituitary gland in the forehead. It has the shape of a hollow cone which is composed of the misty white energy of hyperspace. Light coming into this vortex is then transferred by a light cord to a visual energy module which is located in a co-dimension of hyperspace. Because these modules are interconnected by light cords, the mind module is able to interpret the visual pattern the eye is seeing. More importantly, the mind can give logical instructions to this spiritual eye module for it to rotate around or move in a particular direction.

- [0013]

  The reason that hyperspace has a white misty look to it is that the speed of light is very much less than the speed of light in our spacetime. The Lorentz transformation says that the distance L is shortened relativistically to a distance L' in a way related to the ratio of the velocity v of the object to the velocity of light c. $L' = L \sqrt{1 - v^2 c^2}$ If the velocity of light is very low, then a small velocity creates an enormous contraction in length. By moving through hyperspace, therefore, enormous distances can be traversed. And this is the reason that it is possible to project the spiritual eye, voice and hearing to remote locations in the galaxy.

- [0014]

  The universe is composed of subspace, space and hyperspace which are co-dimensions of each other. Subspace is defined by the geometry of the tetrahedron which is a four-sided solid whose faces are equilateral triangles having three 60° angles. Referring to FIG. 5, a tetrahedron whose sides are the square root of three (A), has a height of the square root of two (B), and base length equal to the square root of one (C). This forms the basic number set {√{square root over (1)}, √{square root over (2)}, √{square root over (3)}}.

- [0015]

  Referring to FIG. 6, the tetrahedron (A) is circumscribed by the sphere (B). Rod (C) is the sphere radius. A second rod (D), of equal length to rod (C), from the center of the sphere to the corner of the tetrahedron makes an angle φ

  of $\phi = \text{ArcSin} (1\ 3) = 19.47122063$ °

  So the four corners of the tetrahedron touch the sphere.

- [0016]

  This tetrahedral geometry can be seen throughout the planets of the solar system Referring to FIG. 7, the islands of the Caribbean curve down from Puerto Rico to Venezuela forming an island vortex. The low density hyperspace energy releasing from the corner of the tetrahedron softens the rock mantle. The hot magma then rises through the rock with the least resistance. This creates a circular arc of volcanic islands along the edge of the vortex.

- [0017]

  Referring to FIG. 8, the Giant Red Spot of Jupiter is located at a southern latitude of 19.5°. This vortex is so large that the entire earth can fit in it.

- [0018]

  Referring to FIG. 9, the Olympus Mons volcano is located at a northern latitude of 19.5° as shown by the marker. This volcano is the size of France. Notice the fallen plume of volcanic debris toward the north east.

- [0019]

  The double harmonic of the tetrahedral angle is twice 19.5° or 39° which is the location of the Silver Bridge in Point Pleasant, West Va. A large wormhole opened up around the bridge during Christmas rush hour when the bridge was full of cars. Due to the low density hyperspace energy, the rivets holding the cables down popped loose and all the cars were dumped into the river. A computer simulation using Schrodinger's quantum mechanics equation for a particle in a potential well shows that as the energy becomes less dense, the particle is no longer contained in the potential well. The electron jumps out. Thus the atomic bonds are broken which softens the rivets. This is the first time that there has been an understanding of the failure mechanism of this bridge.

- [0020]

After downloading from the Internet several pages of the index of refraction of a wide range of materials, I noticed that the index of refraction for Plexiglas was 1.50. Another source said it was 1.51. One of the Internet sites had a movable flashlight which showed the incident ray and the refracted ray. For Plexiglas, surprisingly enough, the incident ray was coming in at an angle of 60° to the normal, and the light was refracted at 35.26°, both of which are tetrahedral angles. The angle of the equilateral face of the tetrahedron is of course 60°. The angle at the top of tetrahedron is the arc-cosine of the ratio of the height over the edge length. $\theta = ArcCos \left[\sqrt{(2 \ 3)} \right] = 35.26°$

According to Snell's law, the index of refraction n, times the sine of the angle $\sin(\theta_1)$ of the ray leaving material m_1, is equal to the index of refraction n_2 times the sine of the angle of refraction $\sin(\theta_2)$ of the ray entering material m_2. Referring to FIG. 10, the equation is

$n_1 \sin(\theta_1) = n_2 \sin(\theta_2)$

The index of air n_1 is equal to one. The index n_2 of Plexiglas is 1.50. If the incident ray is at $\theta_1 = 60°$ to the normal, then the output angle

is $\theta 2 = ArcSin \left[n 1 n 2 \ \times Sin \left[(\theta 1) \right] \right] = ArcSin \left[2 3 \ \times 3 2 \right] = 35.26°$

which is equal to the angle of the tetrahedron.

- [0021]

Then I recalled several months earlier that I had gone to the Subway restaurant to get a sandwich. I was sitting by the Plexiglas window communicating with the Admiral whose mothership was in earth orbit. She mentioned that they were bringing two people aboard. At that moment I looked out through the window and I could see both of them clearly and easily through my spiritual eye. To my amazement, I saw both of the captives start to pull out guns from behind their backs. I then projected by spiritual hands which resulted in preventing the attack on the security guards. To say the least, the Admiral was rather surprised at these events. She then asked me to look at the design of the window because she thought it had something to do with my enhanced remote viewing capabilities. Looking at the Plexiglas, I noticed that on the edge of the large window pane there was a shorter piece of Plexiglas which was mounted parallel to the window pane. This smaller panel acted to protect the yellow neon fluorescent tube. I took the measurements of the design using a piece of paper that I found near the table.

- [0022]

I then went home and designed up a mounting bracket with my 3D computer software. I had already installed the stereolithography software that converts the design to the *.STL file format. How stereolithography works is that it slices the design into many thin horizontal sections. The machine has a platform which is mounted in a bath of liquid polymer. An ultraviolet laser, mounted on an xy-table, then traces out the slice. Because the liquid polymer is light sensitive, it polymerizes immediately into solid plastic. Then the platform is lowered a few thousandths of an inch and the second slice is added. This process eventually builds up the complete 3D part. Using the Internet, the *.STL file is sent by e-mail to the stereolithography service provider who returns the part overnight. So the next day I had the piece from which I made a plastic mold and several additional pieces for mounting the Plexiglas bracket on a full sheet of Plexiglas that I ordered locally. It never occurred to me to measure the angles, so after I got the index of refraction for Plexiglas, I measured the incident angle and it turned out to be 61°. So then I realized that the spiritual eye was being diffracted across these two Plexiglas plates which connected it to the tetrahedral geometry of subspace. Because subspace is the foundation of space, this created a much more efficient route for remote viewing. The result is that this invention has allowed me to make contact with 430 alien civilizations. Since then I have been awarded the Aphysics prize for my work in the invention and elaboration of the tetrahedron diagram of which there are now over 4000 graphs. The scientific discoveries contained in the diagram are (1) the electron and proton are one and same particle, (2) the existence of hyperspace, (3) how mass can be taken out of dimension, (4) cosmology determines the elementary particles, (4) all the physics constants and the tetrahedral geometry are contained in the two 360° circles of the infinity symbol and (5) all the physics constants are determined geometrically and projected from a subspace manifold into our dimension. For my work in Revelations, I was awarded four beautiful galloping riderless white horses of the Apocalypse.

A BRIEF DESCRIPTION OF THE DRAWINGS

- [0023]

  FIG. 1. Remote viewing the top of a palm tree containing several coconuts.

- [0024]

  FIG. 2. Remote viewing three soldiers running across a wooden bridge.

- [0025]

  FIG. 3. Remote viewing computer building that soldiers were guarding.

- [0026]

  FIG. 4. Spiritual eye of human aura.

- [0027]

  FIG. 5. Tetrahedron.

- [0028]

  FIG. 6. Tetrahedron circumscribed by sphere.

- [0029]

FIG. 7. Caribbean volcanic island vortex.

- [0030]

FIG. 8. Giant Red Spot on Jupiter.

- [0031]

FIG. 9. Olympus Mons volcano on Mars.

- [0032]

FIG. 10. Snell's Law of Refraction.

- [0033]

FIG. 11. Perspective view of remote viewing station.

- [0034]

FIG. 12. Wedge-shaped spacers for diffraction panel.

- [0035]

FIG. 13. Remote viewing angle.

- [0036]

FIG. 14. A remote viewing session taking place on a planet located 90,000 light years from earth on the other side of the galaxy showing His Highness signing with his left hand the document that allows his planet to become a member of the Pleiadian Federation.

DETAILED DESCRIPTION OF THE INVENTION

- [0037]

1. Referring to FIG. 11, the remote viewing station is a rectangular box wooden frame (A) on which is mounted on one side a large sheet of Plexiglas (B). The diffraction panel (C), made of a shorter length of Plexiglas, is mounted with acrylic glue on the Plexiglas sheet using clear polyoptic molded plastic spacers (D).

- [0038]

Referring to FIG. 12, the wedge-shaped spacers (D) hold the diffraction panel (C) to the sheet of Plexiglas (B). The angle of the wedge is 30° which makes the incident angle 60° to the normal.

- [0039]

Referring to the top view perspective FIG. 13, sitting on the wide bench, the remote viewer can adjust his sight along the wedge for proper alignment at an angle of 60°. Due to the 1.50 index of refraction of Plexiglas, the spiritual eye is diffracted across the edge of the first panel and then refracted across the second panel at the tetrahedral angle of 35.26°.

- [0040]

    Referring to FIG. 14, the remote viewing image is seen superimposed on the large sheet of Plexiglas which acts as the viewing screen.

Claims

1. A remote viewing station comprising:

(a) a rectangular box frame made of wood having a length of six feet, a width of four feet and a height of six feet;

(b) a large sheet of quarter inch Plexiglas, having an index of refraction of 1.50, mounted on the right side of item (1a);

(c) several wedge-shaped clear plastic spacers, three inches in width and eight and a half inches in length, having a wedge angle of 30° that are mounted on the interior right side of item (1b);

(d) a shorter sheet of quarter inch Plexiglas, having a length of one and a half feet, mounted on item (1c) parallel to item (1b);

(e) a wide bench on which the remote viewer sits so that the remote viewer can align his sight along the wedge angle of item (1c);

2. A remote viewing amplifier that:

(a) diffracts the spiritual eye of the remote viewer across the edge of item (1d) at an incident angle of 60° to the normal;

(b) refracts the spiritual eye at 35.26° to the normal due to the refractive index of item (1a) according to Snell's Law of Refraction; and

(c) aligns the spiritual eye with the tetrahedral geometry of subspace due to the diffraction/refraction combination of item (2a) and (2b).

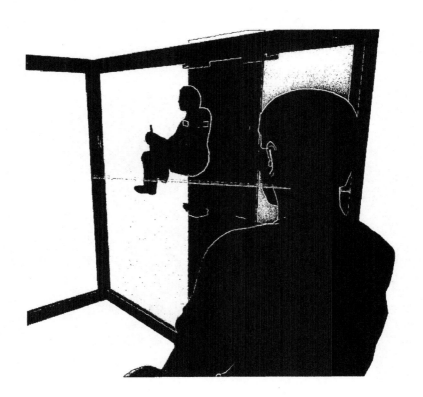

Figure 1

Figure 2

Figure 3

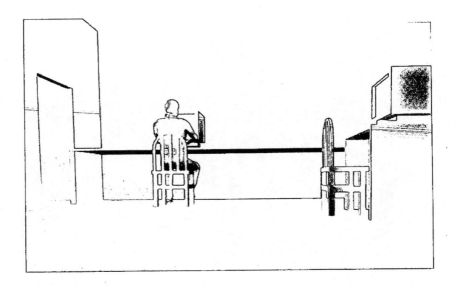

Figure 4

Figure 5

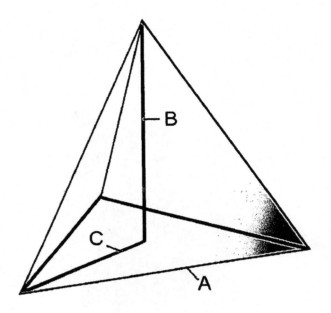

Figure 6

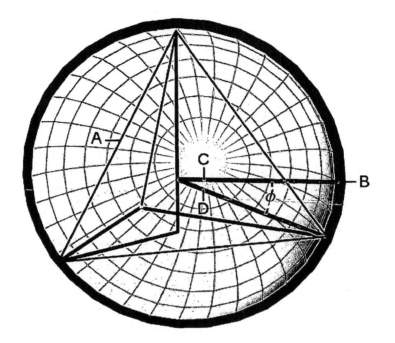

Figure 7

Figure 8

Figure 9

Figure 10

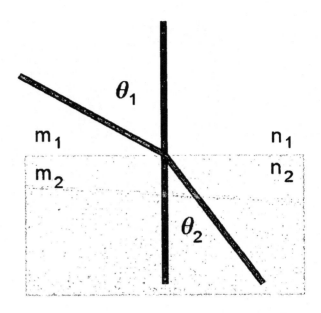

Figure 11

Figure 12

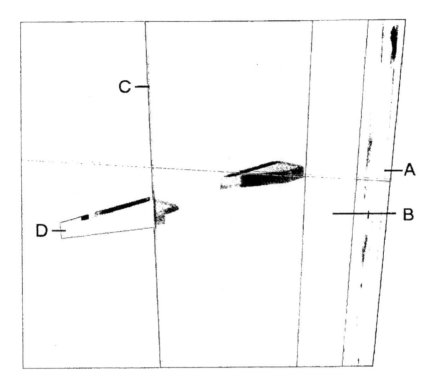

Figure 13

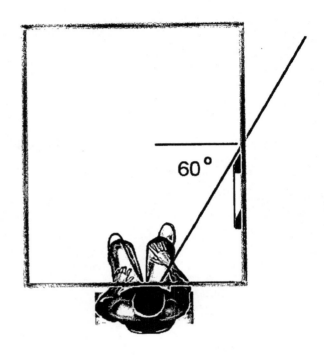

60°

Figure 14

Chapter 7: Electric Dipole Moment Propulsion System

Abstract

A spacecraft propulsion system utilizing a rotating octagon of trapezoidal electrically charged flat panels to create an electric dipole moment that generates lift on the hull.

Classifications

B64G1/409 Unconventional spacecraft propulsion systems

US20060070371A1
United States

Inventor

John St. Clair

Worldwide applications
2004 US

Application US10/958,436 events

2004-10-05

Application filed by St Clair John Q

2004-10-05

Priority to US10/958,436

2006-04-06

Publication of US20060070371A1

Status

Abandoned

Description

BRIEF SUMMARY OF THE INVENTION

- [0001]

  The invention is a spacecraft utilizing trapezoidal electrostatically charged flat plate panels which form a pyramidal hull. A panel contains three holes each of which produces a potential energy ellipsoidal bubble that creates an electric dipole moment. The rotation of the hull generates a magnetic moment and a magnetic field gradient in the vertical direction that produces a lift force on the spacecraft.

  BACKGROUND OF THE INVENTION

- [0002]

  It is known from electrodynamics that a hole in a conducting plane forms a potential energy bubble. This bubble creates an electric dipole moment from which it is possible to develop a magnetic moment. A rotating tilted hull produces a velocity gradient that generates a magnetic field gradient in the vertical direction. This combination produces a lift force on the spacecraft. A very large potential energy bubble is produced provided that the hole protrudes out of the plate in an ellipsoidal shape. Furthermore, a double cladding, in which each layer around the hole has a different permittivity, confines the field to the outside of the hull for even better results.

- [0003]

  The planar potential energy is created by a grid of electrically charged wires or rods running the length of each panel. A circular potential energy from each rod very quickly sums to form a flat sheet of energy which emerges from the hole to form the potential energy bubble.

  SUMMARY OF THE INVENTION

- [0004]

  The invention relates to a spacecraft utilizing a rotating octagon of trapezoidal electrically charged flat plate panels to form a hull in the shape of a pyramid. Each panel has three protruding ellipsoidal bubbles that produce an electric dipole moment from a planar potential energy field created by a group of charged rods parallel to the panel. Because the panels are tilted and the hull is rotating, there is a tangential velocity gradient in the vertical direction. This creates the magnetic moment. Because the hull rotates, the radial electric field produces a magnetic field gradient in the vertical direction. This combination of magnetic moment and magnetic field gradient produces a lift force on the hull of the spacecraft.

- [0005]

  On the underside of each panel is a group of high voltage electrically charged rods which run parallel to the panel. These wires or rods produce a planar electrical potential field underneath the holes in the panel. This potential energy field then bubbles out of the holes in the panel to create a large ellipsoidal potential energy field above the hull. The potential energy bubble carries an electric dipole moment which when rotated with the hull generates a magnetic moment in the vertical direction.

  A BRIEF DESCRIPTION OF THE DRAWINGS

- [0006]

  FIG. 1. Perspective view of electric dipole moment spacecraft.

- [0007]

  FIG. 2. Perspective exploded view of one panel with the ellipsoidal domes, flat hull panel with three holes, the charged rod grid and the planar potential energy field.

- [0008]

  FIG. 3. Planar view of flat potential energy field produced by electrically charge wire rods.

- [0009]

  FIG. 4. Perspective view of cylindrical coordinates {r, θ, z}.

- [0010]

  FIG. 5. Perspective view of ellipsoidal potential energy field emerging from hole in plate which produces an electric dipole moment.

- [0011]

  FIG. 6. Planar view of field lines of potential energy bubble emerging from plate hole.

- [0012]

  FIG. 7. Planar view of sloping hull profile needed to get a velocity gradient.

- [0013]

  FIG. 8. Perspective view of hull showing electric dipole moment, the tangential velocity of the hull, and the magnetic moment.

- [0014]

  FIG. 9. Perspective view of the magnetic moment components in the radial and vertical direction whereby the lift force is generated by the dot product of the vertical magnetic moment with the magnetic field gradient.

- [0015]

  FIG. 10. Perspective view of cross section of dome showing two layer cladding with different permittivities to enhance potential energy field.

- [0016]

  FIG. 11. Planar view of enhanced potential energy field with two layer cladding.

  DETAILED DESCRIPTION OF THE INVENTION

-

  o 1. Referring to FIG. 1, the spacecraft is a rotating octagon of trapezoidal electrostatically charged flat panels which form a closed hull (A). Each panel has three ellipsoidal domes (B) of varying size centrally located along the major length of the panel. The purpose of the dome is to create a large ellipsoidal potential energy bubble over the hull which develops an electrical dipole moment. Because the hull is rotating, a magnetic moment is created in the vertical direction. A magnetic field gradient created by the rotating electric field on the hull in combination with the magnetic moment produces a lift force on the hull.

○ 2. Referring to FIG. 2, the trapezoidal hull panel (A) contains three ellipsoidal holes (E). A group of wires or rods (C) running parallel to and just underneath the panel are electrically charged to a high voltage at the end terminals (B). The rods produce a planar potential energy field (D) just under the holes in the panel. The field emerges from the holes in the shape of an ellipsoidal bubble and is amplified by an ellipsoidal dome (F) on the outside of the hull.

○ 3. Referring to FIG. 3, the group of parallel rods (A) are given a linear charge λ in units of charge per meter. The electric field E developed by the rod is the linear charge divided by the circumference of a circle of radius r around the wire times the permittivity ε of space. The analysis of this arrangement shows that within a few grid width spacings, the potential energy field φ due to the electric field E_0 has become planar (B) in the z-direction given by the equation
$\varphi = -E_0 z$

○ 4. Referring to FIG. 4, the following analysis is done in cylindrical coordinates {r,θ,z}.

○ 5. Referring to FIG. 5, the ellipsoidal potential energy (B) emerges through the hole in the panel plate (A). In doing so it creates an electrical dipole moment (C) shown by the arrow normal to the hole area.

○ 6. Referring to FIG. 6, the bubble (B) emerges through plate (A).

○ 7. Because the bubble has the shape of an ellipse, the centroid y of the bubble would be four thirds the radius a divided by π as given by $y\_ellipse = 4\,3\,$ ▢ $a\,\pi$

○ The electric dipole moment is then given as the charge q times the centroid y. The charge of the hole is equal to the permittivity E times the electric field E emerging from the hole times the area of the hole of radius a▢ $q = \varepsilon\,0$ ▢ E ▢ ▢ ▢ π ▢ ▢ ▢ a 2 ▢ coul 2 m 2 ▢ ▢
▢ newton ▢ newton coul ▢ ▢ ▢ m 2 = coul = charge

○ 8. The electric dipole moment p is the centroid y times the charge
q p = qy = $\varepsilon\,0$ ▢ E ▢ ▢ ▢ π ▢ ▢ ▢ a 2 ▢ 4 3 ▢ a π = 4 3 ▢ $\varepsilon\,0$ ▢ Ea 2 ▢ ▢
▢ coul · meter

○ The electric dipole moment p times the hull velocity v is equal to a magnetic moment μ which is what creates the lift force on the hull▢
$\mu = pv$ amp·m^2

○ 9. The rotating hull creates the electric dipole moment velocity so that the entire hull develops a magnetic moment. In tensor notation, the magnetic moment μ is in the vertical z-direction because there is a radial component of the electric dipole moment times the velocity. The velocity is the radius r in the radial direction times the angular velocity ω in the z-direction

$$\mu^z = p_r \, x \, {}^r\omega^z$$

○ 10. The force F on the hull is the gradient of the dot product of the magnetic moment μ with the magnetic B field

$$F = \nabla(\mu \cdot B)$$

○ 11. By electrically charging the hull of the vehicle, a radial electric field is produced. By rotating the hull, the radial electric field changes with time. Thus Maxwell's equations will involve the curl of the magnetic field in the radial direction because the radial electric field is varying with time $(\nabla \times B)_r = \frac{1}{c^2} \frac{\partial E_r}{\partial t}$

○ 12. The cross product involves the magnetic field in the theta direction which is zero $\frac{1}{r} \frac{\partial B_z}{\partial \theta} - \frac{\partial B_\theta}{\partial z} = \frac{1}{c^2} \frac{\partial E_r}{\partial t} = \frac{1}{r} \frac{\partial B_z}{\partial \theta}$

  ▪ Substituting the derivative of the electric field E

  $$E_r = E_0 \, e^{\,i\omega t}$$

○ and integrating with respect to angle theta gives the vertical magnetic field B as the tangential velocity v times the radial electric field E divided by the speed of light c squared $B_z = \frac{v}{c^2} E_r$

○ 13. The force on the hull is the gradient of the magnetic moment μ times the magnetic field B. In the equation for the magnetic field, the only available variable to work with in order to get a gradient of the magnetic field comes from the velocity. $\frac{d B_z}{dz} = \frac{E_r}{c^2} \frac{dv}{dz}$

○ 14. Referring to FIG. 7, because the hull is in the shape of a pyramid, the velocity is a function of the height z of the hull. Using eight flat sides keeps the radial electric field pointing in the same direction in each panel. Each panel has three domes to produce the magnetic moment for a total of 24 magnetic moment generators.

○ 15. Referring to FIG. 8, the electric dipole moment (A) points in the radial direction, the rotating hull produces a tangential velocity (B), and the result is a magnetic moment (C) along the panel.

○ 16. Referring to FIG. 9, because the magnetic moment is parallel to the panel, there are vertical and radial components of the magnetic moment. The vertical magnetic moment creates the dot product with the magnetic field gradient, which is equal to the lift force.

○ 17. FIG. 10 shows a cross-section of the dome (A) and the plate hole (B) with double cladding to enhance the field. The upper cladding (D) has a low relative permittivity in the range of 2 to 40, and the lower layer has a high relative permittivity in the range of 1200 to 4000.

○ 18. Comparing FIG. 11 to FIG. 7, this dome and cladding configuration creates a much larger electric dipole moment compared to a hole in the plate. The wavy lines are the equi-potential energy lines from the dome (C) and the upper layer (B) and the lower level (A).

Claims

1. A spacecraft propulsion system comprising:
a rotating octagon of trapezoidal electrostatically charged flat panels which form a closed sloping hull in the shape of a pyramid;
panels each having three holes covered by three ellipsoidal domes of varying size centrally located along the major length and axis of each panel; and
a grid of high voltage electrostatically charged rods located on the interior side of each panel such that a planar potential field is produced parallel to and under each panel hole.

2. The domes, holes and rotating charged hull of method 1 producing:
an ellipsoidal potential energy field emerging from the holes and generating an electric dipole moment on the outside of the hull;
a magnetic moment in the vertical direction due to the rotating electric dipole moment; and
a rotating electric field in the radial direction which generates a corresponding magnetic field gradient in the vertical direction proportional to the velocity gradient of the sloping panels of the hull.

3. A lift force on the spacecraft hull generated by:
the magnetic moment times the gradient of the magnetic field in the vertical direction; and
a dual surface layer hull cladding having different permittivities which enhance the electric dipole moment whereby the upper cladding has a low relative permittivity in the range of 2 to 40, and the lower layer has a high relative permittivity in the range of 1200 to 4000.

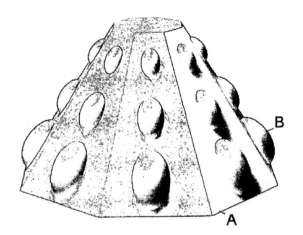

Figure 1

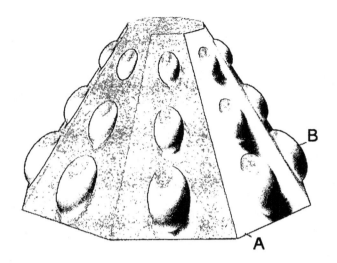

Figure 2

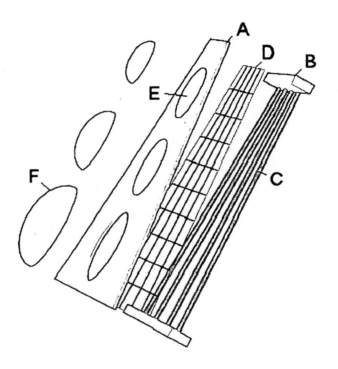

Figure 3

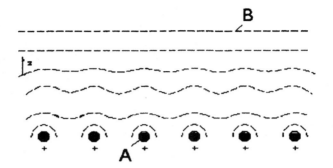

Figure 4

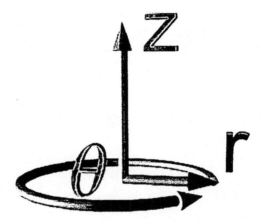

Figure 5

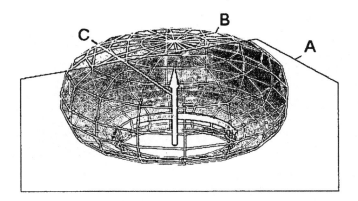

Figure 6

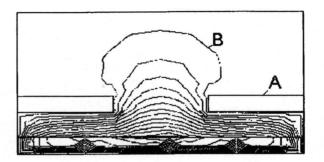

Figure 7

$$\frac{dB}{dz} = \frac{E_r}{c^2} \frac{dv}{dz}$$

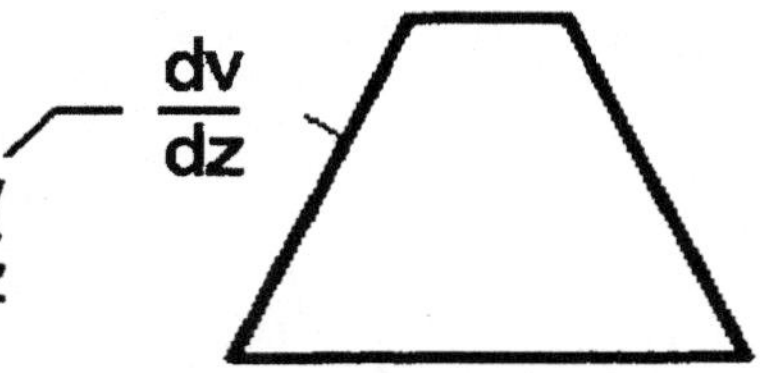

hull profile

Figure 8

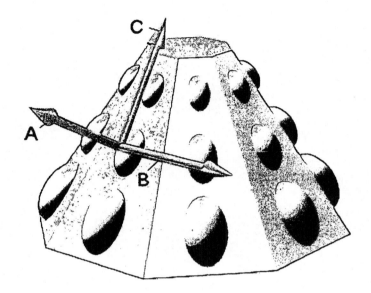

Figure 9

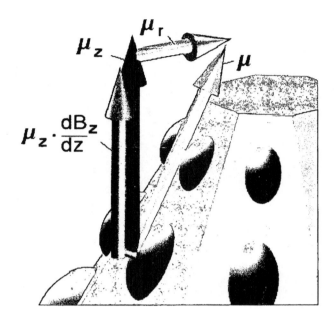

Figure 10

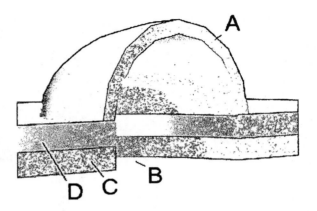

Figure 11

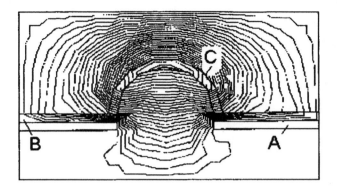

Chapter 8: Permanent Magnet Propulsion System

Abstract

This invention is a propulsion system for a train that uses permanent magnets mounted on a rotating iron cylindrical plate carrying a radial current in order to create a spacetime curvature distortion which pulls the locomotive along the track.

Classifications

H02K99/20 Motors

US20060112848A1
United States

Inventor

John St. Clair

Worldwide applications

2004 US

Application US11/001,217 events

2004-12-01

Application filed by St Clair John Q

2004-12-01

Priority to US11/001,217

2006-06-01

Publication of US20060112848A1

Status

Abandoned

Description

BRIEF SUMMARY OF THE INVENTION

- [0001]

  This invention is a propulsion system for a train that utilizes spinning cylindrical magnets in order to create a spacetime pressure distortion ahead of the vehicle that pulls the locomotive along the track.

  BACKGROUND OF THE INVENTION

- [0002]

  At the present time, referring to FIG. 1, proposed permanent magnet propulsion systems use a dual railway track (A) supporting a series of coil windings (B) located along the track. The vehicle is attached to two permanent magnets (D) between steel pole pieces (C). The north pole of each magnet faces the interior pole piece such that the magnetic flux path (E) follows the center pole piece up through the railway bed and then back to the south pole of the magnet. As the magnets move along the track, the coil windings are activated at the correct time by Hall sensors. With the coil energized as a north pole, the permanent magnet north pole is repelled which drives the vehicle along the track The problem with this design, and other similar designs, is that it is not practical to wind huge numbers of sensor-activated electrical coils along a steel track.

- [0003]

  From Einstein's General Theory of Relativity, it is known that a spacetime curvature pressure develops perpendicular to direction of vibration of the electric and magnetic field. As an example, the photon has an electric field vibrating in the vertical y-direction and a magnetic field vibrating in the horizontal x-direction. The spacetime curvature pressure is therefore along the z-axis of radiation which pushes the negative mass of the photon along. Thus in order to create a spacetime curvature pressure in the z-direction along the track which would pull the train forward, a magnetic flux density field is required in the radial direction.

- [0004]

  Referring to FIG. 2, four equally-spaced north permanent magnets (B) surrounding a centrally-located south permanent magnet (C) are mounted on an iron cylinder which acts as the radial flux return path. The magnetic flux density field (D) is in the radial direction from the north pole to the south pole. In order to provide strength, the magnets are molded onto a steel shaft and coated with epoxy so that they don't rust. During the molding process, a capacitor-discharge magnetizer is used to create the magnetic field of the magnet.

- [0005]

  In Cartesian coordinates $\{-ct,x,y,z\}$, the elemental spacetime length ds squared is the sum of the squares of the incremental lengths $\{cdt,dx,dy,dz\}$
  $$(ds)^2 = -(dt)^2 + (dx)^2 + (dy)^2 + (dz)^2$$
  where the speed of light c is unity. The coefficients $(-1,1,1,1)$ of this equation make up the g metric 4×4 tensor ▢ ▢ [t ▢ ▢ ▢ x ▢ ▢ ▢ y ▢ ▢ ▢ z] g αβ = t - 1 ▢ 0 0 0 x 0 ▢ 1 0 0 y 0 0 1 0 z 0 0 0 1

- [0006]

The Faraday electromagnetic tensor contains the magnetic fields which determine how the spacetime length ds is curved. For a magnetic flux density field in the x-direction, Bx, and a magnetic flux density field in the y-direction, By, the Faraday tensor is ⏃ ⏃ t ⏃ ⏃ ⏃ x ⏃ ⏃ ⏃ y ⏃ ⏃ ⏃ z F β α = t 0 0 0 0 x 0 0 0 -
By y 0 0 0 Bx z 0 By - Bx 0

The stress-energy-momentum tensor T, which determines how space is curved, is calculated from the following equation 4 ⏃ π ⏃ ⏃ ⏃ T μ ⏃ ⏃ ⏃ v = F μα ⏃ F α v - 1 4 ⏃ g μ ⏃ ⏃ ⏃ v ⏃ F αβ ⏃ F αβ

The stress-energy in the z-direction ahead of the locomotive
is $T_{zz} = B x 2 + B y 2 8 ⏃ π = B r 2 8 ⏃ π$
where the sum of the squares of the fields in the x and y directions is the radial B field. In Einstein's General Relativity Theory, the curvature G tensor is equal to the stress-energy tensor divided by 8π. The G tensor is the curvature of space having units of inverse radius squared. $G = T 8 ⏃ π$

Therefore the curvature G_{zz} generated along the z-direction ahead of the train is proportional to the square of the magnetic flux density field G zz = 1 r 2 = G ⏃ ⏃

⏃ ε c 2 ⏃ B r 2 8 ⏃ π = 1 meter 2

where G is Newton's gravitational constant (not to be confused with the curvature tensor), ε is the linear capacitance of space, and c is the speed of light. The linear mass of space Ω is the speed of light c squared divided by the gravitational constant G, so that the equation can be written as G ⏃ ⏃ ⏃ ε c 2 ⏃ B r 2 8 ⏃ π = ⏃

⏃ ε Ω ⏃ B r 2 8 ⏃ π = 1 Ω ε ⏃ B r 2 8 ⏃ π

where the conversion factor is the square of the magnetic vector potential

A Ω ε = kg ⏃ ⏃ ⏃ m sec ⏃ ⏃ ⏃ coul = A

which is actually the momentum per charge. Therefore the curvature equation can be written as 1 r 2 = 1 8 ⏃ π ⏃ (B r A) 2

This equation shows that it is necessary to create a magnetic vector potential together with the radial magnetic flux density field in order to create a curvature of space. Looking at the units of A shows that it is a mass momentum per charge A = kg sec ⏃ m coul = m ⏃ ⏃ ⏃ ω 2 ⏃ r I

or a mass m rotating with angular velocity c) per current along the radius. In terms of the invention, what this means is that the mass of the iron cylinder has to be rotating and there has to be a radial electrical current I in order to produce the linear charge along the radius. The differential mass dm depends on the circumference times the differential radius dr, the mass density p, and the length L of the cylinder
dm=ρ2πrLdr
so that the magnetic vector potential becomes A = ∫ 0 R ⏃ ρ ⏃ ⏃ ⏃ 2 ⏃ π ⏃ ⏃ ⏃ rL ⏃ ⏃ ⏃ ω 2 ⏃ r I ⏃ ⏃ ⏃ d r = 2 3 ⏃ R 3 ⏃ ρ ⏃ ⏃ ⏃ π ⏃ ⏃ ⏃ L ⏃ ω 2 I

The value of A for the iron cylinder
is L = .2 ⏃ m ρ = 7866 ⏃ kg m 3 R = 1 ⏃ m ω = 2 ⏃ ⏃ ⏃ π ⏃ ⏃ ⏃ f = 6.28 ⏃ ⏃ ⏃ sec - 1 I = 3000000 ⏃ ⏃ ⏃ amp A = .04335 ⏃ kg ⏃ ⏃ ⏃ m sec ⏃ ⏃ ⏃ coul Br = 1.2 ⏃ tesla 1 8 ⏃ ⏃ ⏃ π ⏃ (Br A) 2 = 30.47 ⏃ m 2 r curvature = 8 ⏃ ⏃ ⏃ π ⏃ (A Br) = .181 ⏃ m

What makes this possible is that the new N-machines can easily generate a minimum of 6 million amps which is twice the value of the electrical current above.

- [0007]

Referring to FIG. 3, the assembly consists of a large induction motor (A) mounted on the train's base plate (B) driving a motor shaft (C) attached to the iron cylinder (D). The shaft is held in place by two thrust bearings mounted in two pillow blocks (E,F). The current-generating N-machine (G) is electrically connected by a copper bus (H) to a copper-beryllium brush (I) on the motor shaft with a similar return brush (J) on the edge of the iron cylinder. The current (K) flows through the motor shaft to the center of the rotating cylinder and then radially outward to the edge. The magnetic flux density flows from the north poles of the outer permanent magnets to the central south pole, along the central magnet to the center of the rotating cylinder and then radially outward to the south poles of the outer magnets.

- [0008]

The thrust F developed is the radius of curvature of spacetime r_c calculated above times the magnet flux density field times the current I $F = r c \boxtimes B r \boxtimes I 8 \boxtimes \boxtimes$
$\boxtimes \pi \approx 30000 \boxtimes$ lbf
Using conservation of tensor coordinates, the radius of curvature is in the z-direction, the magnetic flux density field is in the radial direction and the current is in the radial direction
$F^z = x^z B_r I^r$
where the radial indices cancel, leaving the z-index as the direction of the force.

SUMMARY OF THE INVENTION

- [0009]

It is the object of this invention to create a spacetime curvature in front of a train locomotive in order to pull the vehicle along the track It is known from gravitational physics that a spacetime curvature is generated perpendicular to the direction of vibration of the electric and magnetic field. A radial magnetic field, which can be produced by permanent magnets attached to the flat faces near the rim of a iron cylinder rotating about the z-axis, will create a curvature in the z-direction. Four cylindrical north-pole-oriented magnets produce a radial magnetic flux density with is channeled into a central cylindrical south-pole-oriented magnet. The flux lines then flow radially outward through the steel rotating cylinder and reconnect with the south poles of the four outer magnets. The rotating iron cylinder generates the equivalent of a magnetic vector potential when an electrical current flows from the center of the cylinder to the edge. This current is generated by an N-machine current generator. The square of the magnetic flux density divided by the magnetic vector potential is equal to the spacetime curvature. The square root of the inverse of the spacetime curvature is the radius of curvature. The thrust developed is this radius of curvature times the magnetic flux density field times the current.

A BRIEF DESCRIPTION OF THE DRAWINGS

- [0010]

FIG. 1. Perspective view of proposed permanent magnetic propulsion system using coil windings on the steel track.

- [0011]

FIG. 2. Perspective view of permanent magnet rotor assembly.

- [0012]

FIG. 3. Perspective view of system showing motor drive, N-machine and permanent magnet rotor.

- [0013]

  FIG. 4. Perspective view of locomotive and rotor/magnet assembly.

  DETAILED DESCRIPTION OF THE INVENTION

-

  - 1. The permanent magnets are made of neodymium-iron-boron material which is heated to its melt temperature and injection molded around a steel shaft threaded at one end while at the same time a pulsed magnetic field is applied to the material using a charge-discharge magnetizer. Because of the iron in the material, a coat of epoxy is applied to the magnet in order to protect it from the environment. Holes are drilled into the iron plate 90° apart near the rim, threaded, and then the steel shaft with the magnet is then inserted. Another hole is drilled and tapped in the center of the circular plate for attaching the south pole magnet which is used as the return path for the magnetic flux.

  - 2. Another easier way to make the magnets is to purchase short lengths of tubular NdFeB magnets and then stack them on the steel shaft with a cylindrical iron pole piece on the end of the shaft. The pole piece then holds the magnets down in place when the shaft is threaded into the plate.

  - 3. Referring to FIG. 4, the propulsion system is mounted inside the train cabin such that the rotor/magnet assembly extends out in front of the locomotive where the spacetime curvature is generated.

Claims

1. A train propulsion system consisting of the following components:

a. a rotating iron cylindrical plate rotor of high relative permeability driven by an induction motor and horizontal steel motor shaft mounted in pillow block thrust bearings;

b. four cylindrical magnets, each molded to a steel support shaft threaded into the iron plate at 90° intervals around the rim of the plate with their north poles facing away from the plate;

c. a fifth cylindrical magnet molded to a steel support shaft which is threaded into the center of the iron plate with the south pole facing away from the plate;

d. an N-machine current generator supplying a radial electrical current from the center of the rotating plate by means of a copper-beryllium brush on the motor shaft (**1** *a*) and another similar brush on the outside edge of the rotor.

e. a locomotive train on which the components are mounted such that the rotor/magnet assembly extends out in front of the locomotive with the rotor's angular velocity vector pointing along the track.

2. a closed magnetic flux path along a radial path in air from the north poles of the four outer magnets (**1** *b*) to the south pole of the central magnet (**1** *c*), through the center magnet and then radially outward through the rotor (**1** *a*), returning back through the four outer magnets, such that the flux and electrical current (**1** *d*) flow in the same outward radial direction through the rotor.

3. the creation of a spacetime curvature due to claims (**1** *a* through **2**) that produces a large force on the locomotive equal to the radius of the spacetime curvature times the flux times the current.

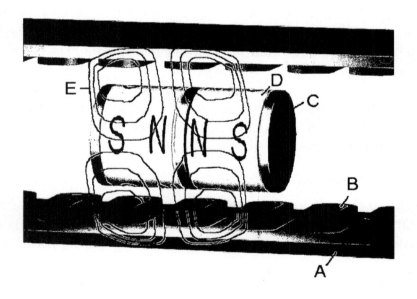

Figure 1

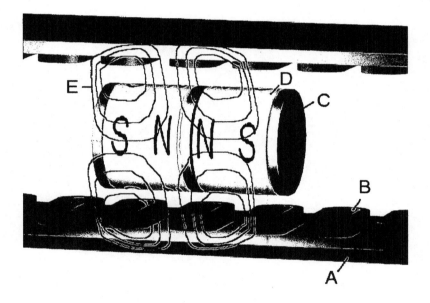

Figure 2

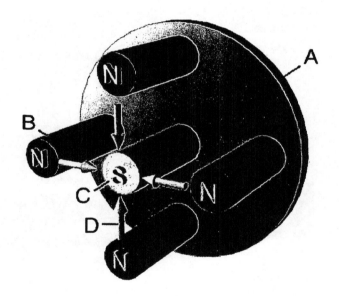

Figure 3

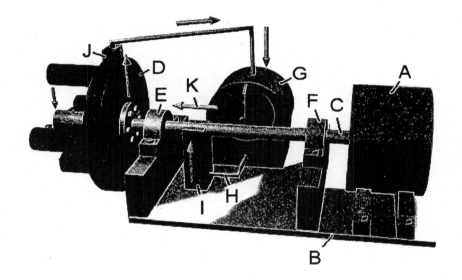

Figure 4

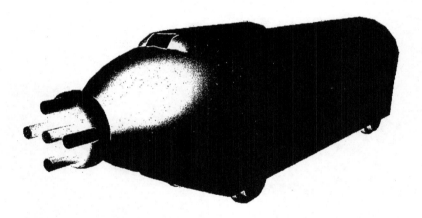

Chapter 9: Triangular spacecraft

Abstract

A spacecraft having a triangular hull with vertical electrostatic line charges on each corner that produce a horizontal electric field parallel to the sides of the hull. This field, interacting with a plane wave emitted by antennas on the side of the hull, generates a force per volume combining both lift and propulsion.

Classifications

B64G1/409 Unconventional spacecraft propulsion systems

US20060145019A1

United States

Inventor

John St. Clair

Worldwide applications

2004 US

Application US11/017,093 events

2004-12-20

Application filed by St Clair John Q

2004-12-20

Priority to US11/017,093

2006-07-06

Publication of US20060145019A1

Status

Abandoned

Description

BRIEF SUMMARY OF THE INVENTION

- [0001]

  This invention is a spacecraft having a triangular hull with vertical electrostatic line charges on each corner. The line charges create a horizontal electric field that, together with a plane wave emitted by antennas on the side of the hull, generates a force per volume providing a unique combination of both lift and propulsion.

BACKGROUND OF THE INVENTION

- [0002]

  Referring to FIG. 1, the spacecraft has a hull in the shape of an equilateral triangle. A parabolic antenna (E) is centrally located in the bottom of the hull. An array of horizontal slot antennas is located along the side of the hull (A). Each back corner (F,G) has a corner conducting plate which is charged to a positive voltage +V. The forward corner (C) has a conducting plate charged to a negative voltage −V. A motion control hemisphere (D) is located on the bottom surface in each of the three corners.

- [0003]

  Referring to FIG. 2, two planes (A,B) intersect at the origin O at an opening angle β. Each plane (x,y) is charged to a voltage V. The potential at point P is determined in polar coordinates {ρφ}. The Laplace equation for the potential Φ in polar coordinates is given by:

  $$\frac{1}{\rho}\frac{\partial}{\partial\rho}\left(\rho\frac{\partial\Phi}{\partial\rho}\right) + \frac{1}{\rho^2}\frac{\partial^2\Phi}{\partial\phi^2} = 0$$

  Using a separation of variables solution, the potential is given as the product of two functions:

  $$\Phi(\rho,\varphi)=R(\rho)\Psi(\varphi)$$

  which when substituted into the Laplace equation becomes:

  $$\frac{\rho}{R}\frac{d}{d\rho}\left(\rho\frac{dR}{d\rho}\right) + \frac{1}{\Psi}\frac{d^2\Psi}{d\phi^2} = 0$$

  Since the two terns are separately functions of ρ and φ respectively, each one has to be constant with the sum of the constants equal to zero:

  $$\frac{\rho}{R}\frac{d}{d\rho}\left(\rho\frac{dR}{d\rho}\right) = v^2 \quad \frac{1}{\Psi}\frac{d^2\Psi}{d\phi^2} = -v^2$$

  These two equations have solutions:

  $$R(\rho)=a\rho^{v+b\rho^{-v}}$$
  $$\psi(\varphi)=A\cos(v\varphi)+B\sin(v\varphi)$$

  The azimuthal angle φ is restricted to a value in the range 0≦φ≦β. The boundary condition is that the potential Φ is equal to V for any radius ρ when φ=0 and φ=β. This means that v has to be an integer value of π so that the sine function is zero:

  $$\sin(v\beta) = \sin\left(\frac{m\pi}{\beta}\beta\right) = \sin(m\pi) = 0 \quad m = 1, 2 \ldots$$

  which in turn means that the coefficient A of the cosine term has to be zero in the solution above. Choosing b=0 makes the general solution for the potential equal to:

  $$\Phi(\rho,\phi) = V + \sum_{m=1}^{\infty} a_m \rho^{m\pi/\beta} \sin\left(\frac{m\pi\phi}{\beta}\right)$$

  which shows that when the angle is zero, the sine is zero and the potential is V. If the angle is β, then there is a multiple of π such that the sine is zero again.

- [0004]

Because the series involves positive powers of the radius, for small enough ρ, only the first term $m=1$ in the series is important. Thus around ρ=0, the potential is approximately

$\varphi(\rho,\varphi) \approx V + a_1 \rho^{\pi/\beta} \sin(\pi\varphi/\beta)$

- [0005]

The electric field component is the negative gradient of the potential:

$E\phi\ (\rho,\phi) = -\dfrac{1}{\rho}\dfrac{\partial\Phi}{\partial\phi} = -\dfrac{\pi}{\beta}\ a_1\beta\ \rho^{(\pi/\beta)-1}\cos(\pi\phi/\beta)$

The surface charge distribution σ at φ=0 and φ=β is equal to the electric field perpendicular to the surface times the permittivity of space ε_0:

$\sigma\ (\rho) = \varepsilon_0\ E\phi\ (\rho,0) = -\varepsilon_0\ \dfrac{\pi}{\beta}\ a_1\beta\ \rho^{\frac{\pi}{\beta}-1}$

Notice that if angle of intersection β is less than π, then the equation says that there is a very small radius to a positive power which means little charge density accumulation.

- [0006]

Referring to FIG. 3, the value of β, in the case of the triangular hull, is equal to 360° less 60° for a total of 300° or:

$\beta = 300\ \dfrac{180}{\pi} = \dfrac{5}{3}\ \pi \quad \rho^{\pi \frac{5}{3}\ \pi - 1} = \dfrac{1}{\rho^{2.5}}$

which says that there is a charge density singularity to the two fifths power for small radius. Thus, the corner plates on the hull create a huge line charge density along the sharp vertical corner edge. The equation for the potential of a line charge density is given as:

$\Phi\ (x,y) = -\dfrac{\lambda}{2\ \pi\varepsilon_0}\ Ln\ ((x-x_0)2 + (y-y_0)2)$

where λ is the charge per unit length in the vertical z-direction, and x_0 and y_0 are the location of the line charge in the xy-plane.

- [0007]

Referring to FIG. 4, the triangular hull (D) is plotted together with the potential contours (A) and the electric field arrows (B) created by the three corner line charges. The line charges are perpendicular to the paper. Notice that the electric field arrows are parallel crossing the center parabolic antenna (C). The electric field is also parallel to the sides (D) of the triangle.

- [0008]

Referring to FIG. 5, along the side of the triangle (A), an array (B) of horizontal slot antennas emit electromagnetic waves that have a vertically polarized electric E field (C). These traveling waves interact with the electric field (D) produced by the line charges on the corners of the triangle.

- [0009]

Using differential forms mathematics, this combination of fields is represented by the Hodge star of the differential of the wedge product of the two fields. The antenna electromagnetic field is a combination of a traveling magnetic field B_w, and electric field E_w. The stationary field E created by the line charges is perpendicular to the traveling wave.

$* d\ (E \wedge (B_w + E_w \wedge dt))\ \varepsilon\ c = \text{force volume}$

where ε is the linear capacitance of space and c is the speed of light. Thus there is a force per volume around the hull.

- [0010]

This combination of fields produces a spacetime curvature as determined by Einstein's General Theory of Relativity. The traveling electric field has an amplitude in the vertical z-direction and travels in the x-direction

$E_{w=E_z}\cos(x-t)$

The Faraday electromagnetic tensor contains all the electric and magnetic fields in all the {x,y,z} directions. The first row and first column contain the two electric fields

F β α = t x y z ⬚ ⬚ 0 E x 0 E z ⬚ cos ⬚ (x - t) E x 0 0 0 0 0 0 0 E z ⬚ cos ⬚ (x - t) 0 0 0 ⬚

The stress exerted on spacetime occurs in the xx, yy and zz-direction as calculated from the stress-energy tensor T of gravitational physics

4 ⬚ π ⬚ ⬚ ⬚ T μ ⬚ ⬚ ⬚ v = F μ ⬚ ⬚ ⬚ α ⬚ F α μ - 1 4 ⬚ g μ ⬚ ⬚ ⬚ v ⬚ F α ⬚ ⬚ ⬚ β ⬚ F αβ

where g is the metric tensor for Cartesian space

g αβ = t x y z ⬚ ⬚ - 1 0 0 0 0 1 0 0 0 0 1 0 0 0 0 1 ⬚

where the diagonal components are the coefficients of the elementary spacetime length ds squared

$(ds)^2 = -(dt)^2 + (dx)^2 + (dy)^2 + (dz)^2$

The calculation produces three stresses T^{xx}, T^{yy} and T^{zz} in their respective {x,y,z} directions.

- [0011]

Referring to FIG. 6, these three stresses are plotted together as a **3D** vector field animated over time in nine frames. The graphs show that there is a lift force as depicted by the vertical arrows as well as a force of propulsion as shown by the interspersed horizontal arrows. With the passage of time, these vectors exchange places with each other so that the lift becomes the propulsion and vice versa, creating a wavy stress-energy field around the hull.

SUMMARY OF THE INVENTION

- [0012]

This invention is a spacecraft with a triangular hull having charged flat plates on the vertical corners of the three sides. The two rear corners are charged to a potential V. The forward corner is charged to a potential −V. The 60° angle on the corner creates a line charge density singularity that produces a huge horizontal electric field pointing from the back to the front of the craft which is also parallel to the sides of the triangle. An array of horizontal slot antennas located on the sides of the triangular hull produce an electromagnetic wave with the electric field polarized in the vertical direction. This combination of fields produces a spacetime force in both the vertical and horizontal directions such that the spacecraft receives a lift force and a force of propulsion.

A BRIEF DESCRIPTION OF THE DRAWINGS

- [0013]

FIG. 1. Perspective view of triangular spacecraft.

- [0014]

FIG. 2. Drawing of the intersection of two charged plates in order to calculate the charge density in the corner.

- [0015]

FIG. 3. Perspective view of the corner angle β for the equilateral triangle.

- [0016]

FIG. 4. Planar 2D graph showing the electric field produced by three line charges on the corners of the triangular hull.

- [0017]

FIG. 5. Perspective view of electric field produced by the linear charge interacting with the traveling electromagnetic wave produced by the slot antenna.

- [0018]

FIG. 6. 3D vector animation of the lift and thrust force generated by the fields.

- [0019]

FIG. 7. Perspective view of slot antenna.

DETAILED DESCRIPTION OF THE INVENTION

- [0020]

Referring to FIG. 7, the antenna (A) is made out of sheet copper in which a rectangular horizontal slot (B) has been notched out using a die press and sheet metal fixture. A coaxial cable from the amplifier and frequency generator is attached across the slot by soldering the outer cable (D) to one side of the slot and the inner cable (E) to the other side of the slot. This creates the positive and negative charges across the gap which forms the vertical electric field (F) which radiates out perpendicularly to the copper sheet.

- [0021]

Although the invention has been described with reference to specific embodiments, such as a particular antenna system, those skilled in the art will appreciate that many modifications and variations are possible without departing from the teachings of the invention. All such modifications and variations are intended to be encompassed within the scope of the following claims.

Claims

1. A spacecraft comprised of the following components:

(a) a triangular hull in the form of an equilateral triangle;

(b) two copper plates attached on opposite vertical sides at each of the three corners of the hull (1 *a*) such that a sharp vertical edge is formed where they come together;

(c) an electrostatic generator used to charge the back two copper-cladded corners (1 *b*) to a high positive voltage, and the third forward copper-cladded corner to a high negative voltage;

(d) a horizontal slot antenna array mounted-on the sides of the hull; and

(e) a frequency generator, antenna and coaxial cables to drive the antenna array (1 *d*).

2. To create, by claims (1 *a*, 1 *b*, 1 *c*), an intense vertical line charge at the corners (1 *b*) and a horizontal electric field that that is parallel to the sides of the hull (1 *a*);

3. To create, by claims (1 *d*,1 *e*), an electromagnetic wave with a vertically polarized electric field traveling outward from the side of the hull (1 *a*); and

4. To create, by claims (2,3), an interaction of the electrostatic field (2) with the electromagnetic wave (3) such that a combined spacetime curvature pressure is generated on the hull in the upward and forward direction to produce lift and propulsion respectively.

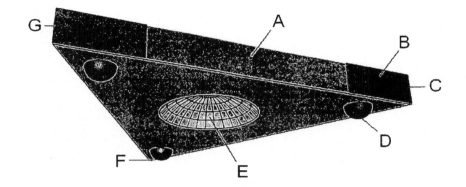

Figure 1

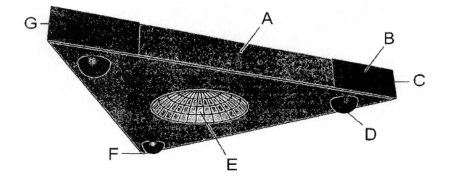

Figure 2

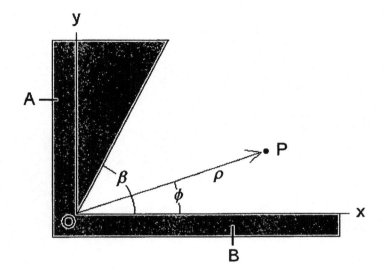

Figure 3

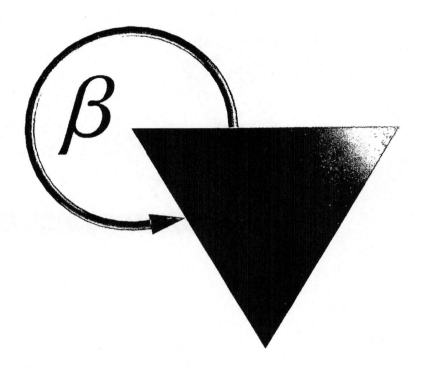

Figure 4

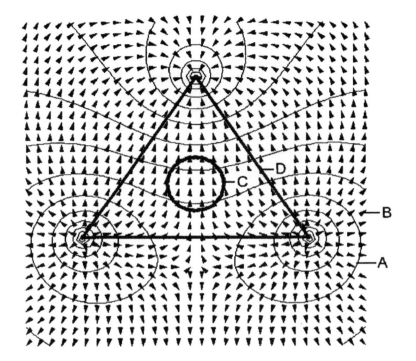

Figure 5

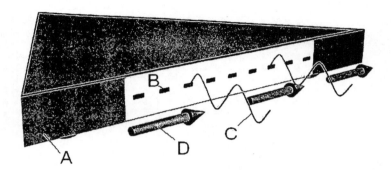

Figure 6

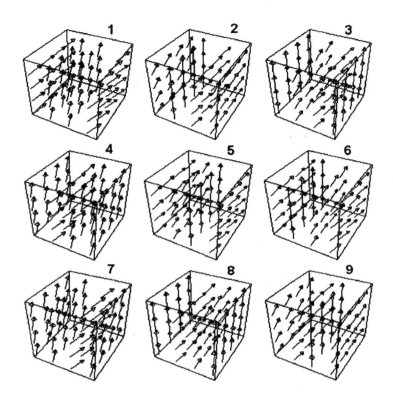

Figure 7

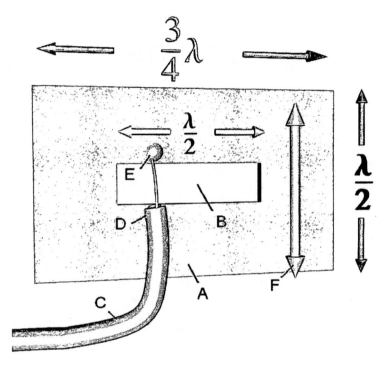

Chapter 10: Photon Spacecraft

Abstract

A spacecraft propulsion system utilizing photon particles to create negative energy over the hull in order to generate a lift force on the hull.

Classifications

F03H99/00 Subject matter not provided for in other groups of this subclass

US20060144035A1
United States

Inventor

John St. Clair

Worldwide applications
2005 US

Application US11/027,969 events

2005-01-03

Application filed by St Clair John Q

2005-01-03

Priority to US11/027,969

2006-07-06

Publication of US20060144035A1

Status

Abandoned

Description

BRIEF SUMMARY OF THE INVENTION

- [0001]

This invention is a spacecraft propulsion system that employs photon particles to generate a field of negative energy in order to produce lift on the hull.

BACKGROUND OF THE INVENTION

- [0002]

Referring to FIG. 1, an electromagnetic wave traveling in the z-direction consists of an electric E field vibrating in the x-direction and a magnetic flux density B field vibrating at right angles in the horizontal y-direction. The energy-stress-momentum of this photon can be analyzed using Einstein's General Theory of Relativity and the Faraday F tensor. The Faraday tensor is a 4×4 matrix containing the electromagnetic wave components as shown here in general where c is the speed of light F β α = t x y z ⧄ □ 0 E x c E y c E z c E x c 0 B z - B y E y c - B z 0 B x E z c B y - B x 0 □
For this particular photon, this tensor is F β α = t x y z ⧄ □ 0 E x c 0 0 E x c 0 0 - B y 0 0 0 0 0 B y 0 0 □

- [0003]

The elemental spacetime length ds squared is equal to sum of the squares of the Cartesian elemental lengths
$(ds)^2 = -(dt)^2 + (dx)^2 + (dy)^2 + (dz)^2$
The coefficients of this equation, {−1,1,1,1} are the diagonal components of the g metric tensor g αβ = t x y z ⧄ □ - 1 0 0 0 0 1 0 0 0 0 1 0 0 0 0 1 □
The stress-energy-momentum tensor T can then be calculated for the photon using the Faraday tensor and the g metric tensor in the following equation from gravitation physics 4 ⧄ π ⧄ □ ⧄ T μ ⧄ □ ⧄ v = F μ ⧄ □ ⧄ α ⧄ F α v - 1 4 ⧄ g μ ⧄ □ ⧄ v ⧄ F α ⧄ □ ⧄ β ⧄ F α ⧄ □ ⧄ β
The stress-energy-momentum tensor T indicates the curvature of space due to the application of electromagnetic fields, mass, angular momentum and charge. The mass of the Earth, for example, generates a negative curvature of spacetime such that objects fall toward the mass. The T tensor, which is also a 4×4 matrix, contains the momentum or flux terms in the first row and first column. The normal pressure stress is located along the diagonal. The shearing stresses are located off the diagonal. The energy term is in the upper left corner as depicted here, T μ ⧄ □ ⧄ v = t x y z ⧄ □ energy flux x flux y flux z - flux x pressure x shear xy shear xz - flux y shear yx pressure y shear yz - flux z shear zx shear zy pressure z □

- [0004]

Since $B^2 = E^2/c^2$, the stress-energy-momentum tensor for the photon is therefore T v μ = t x y z ⧄ □ - E 2 c 2 0 0 + E 2 c 2 0 0 0 0 0 0 - E 2 c 2 0 0 + E 2 c 2 □
This remarkable result shows that the photon is actually a negative energy particle (top left corner) which is pushed along by a positive pressure wave (lower right corner). The particle has a positive flux (upper right corner) in the z-direction, as well as a balancing negative flux in the lower left corner so that the overall momentum of the universe remains the same. All four components cancel and we see the photon as a massless particle moving at the speed of light.

- [0005]

Thus the key idea behind this invention is that it is possible to cancel out the pressure term and leave a stationary vibrating electromagnetic field of negative energy over the hull of the spacecraft. The importance of negative energy is that it is a prerequisite to generating wormholes between space and hyperspace.

- [0006]

Hyperspace consists of the those co-dimensions which have different physics constants such as a low speed of light. The existence of hyperspace, which has a white misty look, is not a well-known scientific concept. Experiments with our magnetic vortex wormhole generators, hyperspace torque generator, full body levitation using Chi Kung breathing, arm levitation by spinning the co-gravitational K field, full body teleportation through hyperspace a distance of 100 meters using a pulsed gravitational wave, jumping into hyperspace, having a plate of toast enfold off the breakfast table and disappear into thin air, walking through walls and doors out-of-dimension, looking into other dimensions, remote viewing through subspace to distances of 100,000 light years, and other electromagnetic experiments carried out by co-researchers, have shown us the reality and existence of hyperspace.

- [0007]

Referring to FIG. 2, the spacecraft consists of an upper (**1**) and lower (**2**) hull attached by ceramic insulators to a circular ring (**3**). The ring provides support and is attached to an outer sharp-edged rim which is electrostatically charged to a potential −V. The purpose of the charged rim is to generate a radial electric E field around the vehicle.

- [0008]

Referring to FIG. 3, the radius of the ring (**4**) is equal to a. The distance from a point on the ring to the z-axis is r. The potential on the z-axis is therefore the charge divided by the distance, $\text{pot} \llbracket \quad \rrbracket Z = q \, a \, 2 + r \, 2$
This potential is expanded as a series in terms of inverse radius r $\text{pot} \llbracket \quad \rrbracket Z \llbracket \quad \rrbracket$
$\llbracket \text{out} = 35 \llbracket q \llbracket \quad \rrbracket \llbracket a \, 8 \, 128 \llbracket r \, 9 - 5 \llbracket qa \, 6 \, 16 \llbracket r \, 7 + 2 \llbracket q \llbracket \quad \rrbracket \llbracket a \, 4 \, 8 \llbracket r \, 5 - qa \, 2 \, 2 \llbracket r \, 3 + q \, r$
The potential outside the ring can be written in terms of the Legendre polynomials
$P \, \text{Vout} = \sum n = 0 \, s \llbracket (a \, r) \, n + 1 \llbracket A \llbracket n \rrbracket \llbracket \text{Legendre} P \llbracket n , \text{Cos} \llbracket (\theta) \rrbracket$
where s is the number of terms in the expansion. By equating the known particular solution potZout on the z-axis with the general Vout solution, the coefficients A[n] are found to be $A \llbracket (0) = q \, a \, A \llbracket (1) = 0 \, A \llbracket (2) = -q \, 2 \llbracket a \, A \llbracket (3) = 0$
which are substituted back into the Vout equation to get the potential outside the ring.

- [0009]

Referring to FIG. 4, the potential (dotted lines **6**) looking at a slice through the ring (**5**) is shown together with the electric E field. The negative gradient of the potential is the electric field (**7**) shown by the direction of the arrows. The importance of this diagram is that the electric field points in the radial direction toward the negatively charged ring. The force on an electron is the electron charge times the electric field
$F = q_e \, E_r = -|q_e|(-|E_r|) = +F$
Because the electron charge is negative and the radial field points in the negative direction toward the ring, the force on the electron is positive. Thus the electron moves away from the ring in the positive radial direction. A 3-dimensional plot of the ring (**8**) and the electric field (**9**) is shown in FIG. 5.

- [0010]

The stress-energy-momentum generated by a radial electric field is calculated using the Faraday F tensor $F\alpha\beta = \begin{bmatrix} 0 & E_r & 0 & 0 \\ E_r & 0 & 0 & 0 \\ 0 & 0 & 0 & 0 \\ 0 & 0 & 0 & 0 \end{bmatrix}$

The g metric tensor has to be given in spherical coordinates $\{r,\theta,\varphi\}$ $g\alpha\beta = \begin{bmatrix} -1 & 0 & 0 & 0 \\ 0 & 1 & 0 & 0 \\ 0 & 0 & r^2 & 0 \\ 0 & 0 & 0 & r^2 \sin(\theta)^2 \end{bmatrix}$

where θ is the angle from the vertical to the radius r. The stress tensor T^{rr} along the radial direction is $T^{rr} = \frac{E_r^2}{8\pi} \quad c^2$

which shows that the pressure is negative along the radial line equal to the square of the radial electric field divided by the square of the speed of light. Because the field is squared, it doesn't matter that the electric field points in the negative direction. The square makes it positive, but the overall curvature pressure is negative. Thus this negative pressure cancels out the positive pressure propelling the photon along. The second key idea of the invention is how to generate this photon moving in the radial direction.

- [0011]

  It has been known for a long time in physics that an electron moving in a circular path will emit photons in a process known by the German word Bremsstrahlung which is translated as "breaking radiation." There are several types of radiation such as classical Bremsstrahlung involving a charged particle making a collision with another charged or uncharged particle in which photons are emitted. The quantum mechanical Bremsstrahlung involves the sudden appearance or disappearance of a charged particle which also emits radiation. In space, having a field of wormholes in which the electrons are spiraling down into hyperspace would result in the emission of photons by the quantum mechanical method. Also, in the atmosphere, having collisions with air molecules results in emission of photons in the classical way.

- [0012]

  In order to get the electrons to spiral around and emit photons, a crossed electromagnetic field is used as shown by the following equation
  $F=q(E_r +v_r \times B_\theta)$
  where the velocity v is in the positive radial direction due to the force of the electric field. The velocity crossed with a magnetic flux density B field in the θ-direction makes the electron move sideways back and forth in a wiggling motion.

- [0013]

  Referring to FIG. 6, a direct current solenoid (10), represented by multiple current loops, running vertically through the center of the hull, generates a magnetic field that curves around the outside of the hull, as shown by contour lines (12). The north pole (11) is at the bottom of the hull. A radial arrow (13) from the electrostatically-charged rim is perpendicular to the magnetic field lines. The cross product in the force equation becomes the electron radial velocity times the magnetic field $v_r B_\theta$.

- [0014]

  Referring to FIG. 7, the electric field is in the y-direction and the magnetic field is in the z-direction. The flat looping path in the x-direction is the motion of the electron. The electron, which has a negative charge, starts to move in the direction opposite to that of the electric field. In this particular diagram, the electron acquires a velocity in the negative y-direction. Then a sideways force in the x-direction is produced due to the cross product of the velocity with the magnetic field times the negative charge
  $-q(-v_y \times B_z)=+F_x$
  Depending on the magnitude of the velocity, various size loops can be produced.

- [0015]

  In terms of the hull coordinates, because the flat loop is in the plane of the electric field which points in the radial direction, the electron emits light in the radial direction. This condition means that the negative radial pressure created by the electric field cancels the radial pressure of the photon. Thus the photon becomes a stationary vibrating quantum of negative energy. This has the appearance of a luminescent light source. The stress tensor for this condition is therefore $T\mu v = \begin{bmatrix} -E2c2 & 0 & 0 \\ E2c2 & 0 & 0 \\ 0 & 0 & 0 & 0 & 0 & -E2c2 & 0 & 0 \end{bmatrix} = -E2c2 \Rightarrow$ residual $\Rightarrow$ negative $\Rightarrow$ energy

  - ○

    - ■ residual negative energy
      which leaves a residual negative energy per photon.

- [0017]

  Referring to FIG. 8, the negatively charged rim (**14**) produces a radial electric field (**16**) that crosses the magnetic B field (**15**) of the solenoid. Electrons emitted by the charged rim then encounter this crossed field which makes them spiral (**17**) around the hull. Because of the tight loop, the electron emits Bremsstahlung radiation in the radial direction (**18**). The positive pressure field of the photon, which is directed in the radial direction, is canceled by the negative pressure field (**19**) created by the electric field. Because the photon energy is negative, a stationary vibrating electromagnetic quantum of negative energy (**20**) surrounds the hull.

- [0018]

  This negative energy and the pressure stress created by the electromagnetic fields open up wormholes between space and hyperspace. The potential head is positive from hyperspace into space because the energy of hyperspace is more positive than the negative energy field. The low-density hyperspace energy fills the hull and its surrounding space with a white misty hyperspace energy which makes the spacecraft lighter in mass, and therefore lighter in weight within a gravitational field. The actual physics is more complicated still because the electrons find that the resistance of hyperspace is lower than the resistance of space. Thus they spiral down the wormholes which results in a sudden disappearance of charge. The quantum mechanical effect of this is to radiate even more photons which in turn produce even more negative energy.

- [0019]

  The lift on the hull is generated by the radial electric field. In cylindrical coordinates, the g metric tensor is $g\alpha\beta = \begin{bmatrix} -1 & 0 & 0 & 0 \\ 0 & 1 & 0 & 0 \\ 0 & 0 & r2 & 0 \\ 0 & 0 & 0 & 1 \end{bmatrix}$
  Using this metric tensor, the pressure stress in the vertical direction T^{zz} is $Tzz = E28\pi \Rightarrow c2$
  which is a positive curvature over the hull. The mass of Earth produces a negative curvature in which objects fall toward the mass. By counteracting this negative curvature with a more than positive curvature, lift is developed on the spacecraft. Because the negative energy lowers the effective mass of the vehicle, the acceleration is large with a modest electric field. Moreover, in our dimension, the speed of light is 299792458 meters per second. Hyperspace energy has a speed of light equal to one meter per second. Thus the stress is amplified by a factor of $A = (299792458 \Rightarrow m/s 1 \Rightarrow m/s)2 \approx 9 \cdot 10^{16}$
  Because electromagnetic fields are relativistic, motion in a low-velocity-of-light energy field amplifies their strength.

SUMMARY OF THE INVENTION

- [0020]

  It is the object of this invention to create a spacecraft propulsion system that produces wormholes between space and hyperspace using negative energy in order to generate lift on the hull. It was discovered in the Riemannian curvature calculations of gravitation physics that negative energy is required to keep open the throat of the wormhole. From experiments with the magnetic vortex wormhole generator, it is known that the proper combination of electromagnetic fields, together with this negative energy, can create a wormhole through which smoke can be blown into hyperspace.

- [0021]

  Referring to FIG. 9, the directions of force, velocity, and electromagnetic fields are referred to in the cylindrical coordinate system $\{r,\theta,z\}$. An electrostatically charged sharp-edged ring in the θ-direction around the hull of the spacecraft produces a radial electric field. A vertical solenoid in the z-direction through the center of the hull produces a magnetic field which is perpendicular at the rim to the electric field. With the current in the solenoid flowing in the clockwise ($-\theta$) direction, using the right-hand rule, the magnetic field points in the upward z-direction outside the rim. Because the rim is charged to a negative voltage, the electric field points toward the hull in the negative radial ($-r$) direction. Electrons emitted by the rim travel outward ($+v$) because the charge on the electron is negative which, together with the negative electric field, produces a positive radial force. The radial force on the electron causes it to acquire a velocity which interacts with the magnetic field. The cross product of the velocity ($+v$) with the positive ($+B$) magnetic field produces a sideways force on the electron in the negative θ-direction. However, because the charge on the electron is negative, the force is
  $F=-q\{v_r,0,0\}\times\{0,0,B_z\}=\{0,qB_z v_r,0\}$
  which is positive in the θ-direction. It is this sideways force that produces a flat spiraling or looping motion whereby the electron emits photons, known in German as Bremsstahlung radiation, in the radial direction. The photon, which is actually a quantum of negative energy, has a positive radial pressure which propels it along. Because the radial electric field produces a negative pressure in the radial direction, the two opposite fields cancel in the radial direction to form a residual stationary vibrating negative energy. Thus the hull becomes surrounded by negative energy which, together with the pressure stresses created by the electric field, generates wormholes between space and hyperspace.

- [0022]

  The gravitational potential between hyperspace and space is positive because the hyperspace energy is more positive than the negative energy around the hull. Thus the low-density, low-speed-of-light hyperspace energy flows through the wormhole and fills the hull. This has the effect of reducing the effective mass of the hull. Because the electric field generates a positive pressure over the hull in the vertical z-direction, there is an upward force on the vehicle due to the pressure times the hull area. Since the vehicle has a low mass, there is a modest upward acceleration on the spacecraft equal to the force divided by mass.

  A BRIEF DESCRIPTION OF THE DRAWINGS

- [0023]

  FIG. 1. Perspective view of an electromagnetic wave.

- [0024]

FIG. 2. Perspective view of spacecraft.

- [0025]

  FIG. 3. Perspective view of charged ring.

- [0026]

  FIG. 4. Planar plot of the radial electric field produced by charged ring.

- [0027]

  FIG. 5. Perspective view of radial electric field around ring.

- [0028]

  FIG. 6. Planar view of magnetic flux density field contour lines.

- [0029]

  FIG. 7. Perspective view of electron motion in crossed electric and magnetic fields.

- [0030]

  FIG. 8. Perspective view of production of negative energy around hull.

- [0031]

  FIG. 9. Perspective view of cylindrical coordinate system {r,θ,z}.

  DETAILED DESCRIPTION OF THE INVENTION

-
  - 1. The hull is made from a single sheet of aluminum which has been stretched to its yield point by hydraulic cylinders. An upper and lower die is CNC machined to the profile of the hull. The soft sheet is then clamped in the die where it takes on the smooth shape of the hull without any wrinkles. The hull is extremely rigid after forming and does not require any structural reinforcements.

  - 2. A section of the aluminum ring is made in a 3D computer graphics program. The model is stored as a stereolithography file (*.stl). The computer model is then sent via Internet e-mail to the stl server who prints the part in an ultraviolet light-cured polymer. The part is returned the next day by Express Mail. Using a rubber blanket mold to create several ring sections, the entire ring is assembled together in another wooden mold box having thin circular laminate-coated particulate wall boards on either side of the ring. Then a liquid rubber mold is poured on top of the ring and allowed to harden overnight at room temperature. Since the rubber mold is flexible, the ring can be extracted fairly easily. This ring model is then sent to the foundry where it is cast in aluminum using the lost wax process in which a wax mold evaporates out of the sand casting. We are also experimenting with non-magnetic copper casting metals containing beryllium having good conductivity.

- 3. A 11.5 cm plastic pipe is mounted on a rotating fixture driven slowly by a microcontoller, stepper motor, and power electronics board. Using a large diameter insulated wire, such as a 17 AWG with a wire diameter of 0.127 cm, the wire is wound slowly on the pipe and expoxied so that the windings don't come loose. The solenoid is then mounted vertically in the hull supported by the support ring and driven by a current generator located nearby on the test rig.

- 4. The ring is driven by a high voltage electrostatic generator similar to the night vision scope high voltage power supplies. The ring charge is isolated from the hull by ceramic insulators.

1. A spacecraft propulsion system comprising the components:

an aluminum horizontal circular structural support ring;

an aluminum hull in the shape of a high dome on top and shallow dome on the bottom attached to the circular support ring using ceramic insulators;

an electrostatically negatively-charged sharp-edged circular ring, preferably of non-magnetic aluminum or copper, attached with ceramic insulators to the outside of the support ring;

a solenoid mounted through the center of the hull in the vertical direction and attached to the center of the support ring;

an electrostatic high-voltage generator to drive the outer electrostatic ring; and

a direct high-current generator to drive the solenoid.

2. The method of claim 1, wherein a negative radial electric field is generated around the hull by placing a negative potential on the sharp-edged electrostatic ring using the electrostatic generator.

3. The method of claim 1, wherein the current-driven solenoid generates a vertical magnetic field around the hull with the north pole of the solenoid facing down through the bottom of the hull which causes the magnetic flux density field to point up outside the rim.

4. The method of claim 1, wherein electrons are emitted radially by the sharp edge of the charged ring.

5. The methods of claims **2**, **3** and **4**, wherein the crossed electromagnetic fields cause the electrons to spiral around in flat loops during which photons are emitted in the radial direction.

6. The methods of claims **2** and **5**, wherein the negative radial pressure created by the electric field cancels the positive radial pressure of the photon to leave a residual quantum of negative energy per photon around the hull.

7. The methods of claims **2** and **6**, wherein the pressure stress created by the electric field, and the negative energy combine to form wormholes between space and hyperspace.

8. The method of claim 7, wherein low-density hyperspace energy of a higher gravitational potential flows through the wormholes to fill the hull and surrounding space around the hull with the effect of reducing the effective mass of the spacecraft.

9. The method of claim 2, wherein the electric field generates a positive pressure in the vertical direction over the hull which together with the hull surface area, generates an upward lift force on the hull.

10. The method of claim 6, wherein the negative energy, having a low light speed, amplifies the strength of the electromagnetic fields and pressure stress fields.

11. The methods of claims **4** and **7**, wherein the electrons spiral down the low resistance wormholes into hyperspace such as to create a sudden disappearance of electrical charge which quantum mechanically causes a large emission of additional photons.

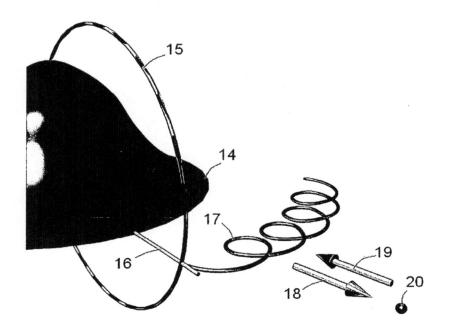

Figure 1

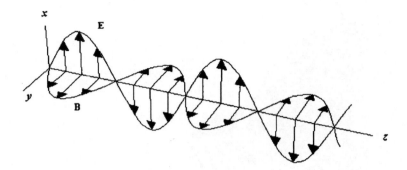

Figure 2

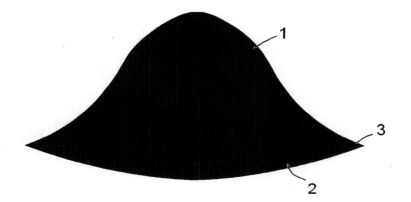

Figure 3

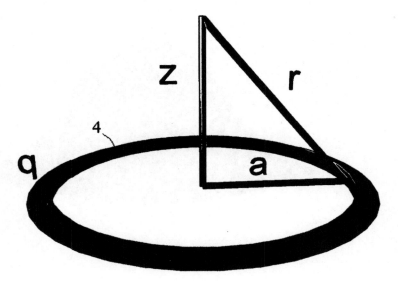

Figure 4

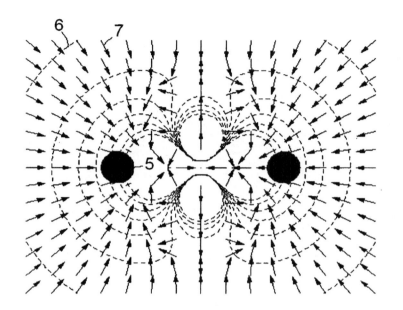

Figure 5

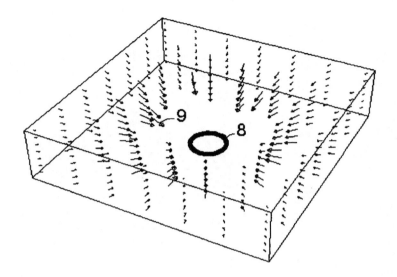

Figure 6

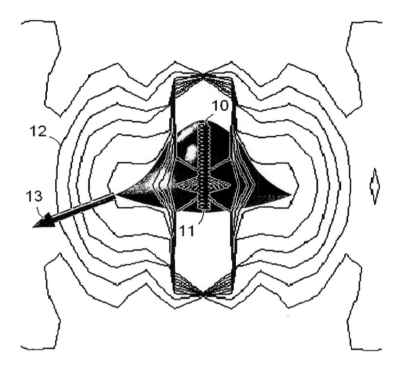

Figure 7

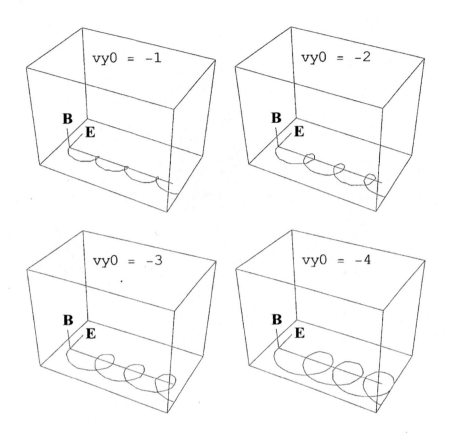

Figure 8

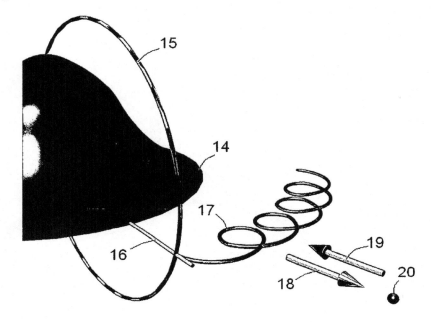

Figure 9

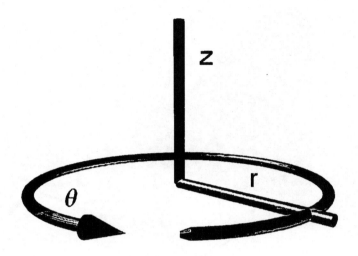

Chapter 11: Water Energy Generator

Abstract

A water energy generator that generates electricity using a magnetic vortex wormhole generator and a water injector/vacuum chamber to produce low density hyperspace energy that causes the hydrogen atoms of water molecules to decay into electron pairs.

Classifications

B64G1/409 Unconventional spacecraft propulsion systems

US20060180473A1
United States

Inventor
John St. Clair

Worldwide applications
2005 US

Application US11/060,037 events

2005-02-17

Application filed by St Clair John Q

2005-02-17

Priority to US11/060,037

2006-08-17

Publication of US20060180473A1

Status

Abandoned

Description

BRIEF SUMMARY OF THE INVENTION

- [0001]

  This invention is an energy generator that uses the transition through wormholes of the hydrogen atoms of water molecules to break the bonds of the atoms and convert the protons into photons and electrons which can be collected for energy.

  BACKGROUND OF THE INVENTION

- [0002]

  A molecule of water consists of two hydrogen H atoms and one atom of oxygen O which has the chemical formula
  H_2O
  The hydrogen atoms can be separated from the oxygen atom by electrolysis. In this process, a direct current of electricity, such as from a battery, is passed through water decomposing it into hydrogen and oxygen. Pure water, however, is a poor conductor of electricity. It is therefore necessary to add some substance to form a solution that will conduct an electric current. Such a solution that will conduct electric current is called an electrolytic solution. A small amount of sulfuric acid or sodium hydroxide is added to the water to form an electrolytic solution. Water electrolyzed yields hydrogen plus oxygen
  Water–>hydrogen+oxygen
  $2H_2O \rightarrow 2H_2 + O_2$
  Because this invention is to be used on spacecraft, the oxygen can be used for breathing and the hydrogen can be used to produce energy that will be used to create the electromagnetic fields which provide lift and propulsion.

- [0003]

  Referring to FIG. 1, a single hydrogen atom consists of one proton (1) in the nucleus and one electron (2) moving in an orbital (3) around the nucleus. In quantum physics notation, there is only one main K shell and one orbital containing a single electron in the 1 s subshell. An orbital is an energy level containing one or two electrons in a subshell of an atom. Only a total of two electrons may be placed in one orbital with the added constraint that the electrons spin in opposite directions. Looking at the 1 s orbital, it can be seen that only half of it is occupied. What this means is that it is possible to add another hydrogen atom in which its electron occupies the other position in the orbital to from the hydrogen molecule H_2.

- [0004]

  In 1925, a physicist by the name of Edwin Schrodinger developed a wave equation, which bears his name, that models the hydrogen atom. Even though the classical picture of FIG. 1 showing a distinct electron orbiting the proton is easy to visualize, in reality the electron is a wavefunction Ψ whose square indicates the probability of finding the electron at a particular point. This then was the start of quantum physics where it was found that the electron energy can only take on certain discrete values.

- [0005]

A traveling wave moving in the positive x-direction can be represented by the function $\Psi_1 = \Psi_1(x,t)$ of the form
$$\Psi_1 = A_1 \cos(2\pi(x/\lambda - vt))$$
where x is the distance along the x-axis, A the wave amplitude, λ the monochromatic wavelength, v the velocity of the wave and t time.

- [0006]

A similar monochromatic wave moving to the left can be represented by
$$\Psi_2 = A_2 \cos(2\pi(x/\lambda + vt))$$
where the sign of the velocity is reversed. The superposition of these traveling waves results in a standing wave, or stationary waves, of the form
$$\Psi = \Psi_1 + \Psi_2 = A\cos(2\pi x/\lambda)\cos(2\pi v t)$$
which is now a product of a spatial-dependent term $A\cos(2\pi x/\lambda)$, and a time-dependent term $\cos(2\pi v t)$. Taking the partial derivative of Ψ twice with respect to x $\frac{\partial^2 \Psi}{\partial x^2} = -\left(\frac{2\pi}{\lambda}\right)^2 \Psi$
The momentum p of a particle is equal to Planck's constant h divided by the mass m of the particle $p = h/\lambda = 2\pi\hbar/\lambda$
where h bar is the reduced Planck constant. Thus Schrodinger's equation can be written as $\frac{\partial^2 \Psi}{\partial x^2} = -\frac{p^2}{\hbar^2}\Psi$
For a particle whose potential energy is V(x), the total energy is the kinetic energy K plus the potential energy
$$E = K + V = (p^2/2m) + V$$
Therefore,
$$p^2 = 2m(E-V)$$
and Schrodinger's equation becomes $\frac{\hbar^2}{2m}\frac{\partial^2 \Psi}{\partial x^2} + (E-V)\Psi = 0$
The potential V is just the Coulomb potential of the product of two charges e divided by the radius r between them $V = \frac{e^2}{4\pi\varepsilon r}$
where ε is the linear capacitance of space. A general wavefunction can be separated into a radial R part and a spherical harmonics part Y
$$\Psi = R_{n,l}\, Y_{l,m}(\theta,\varphi)$$
where the first term is called a radial wavefunction which describes the "in-out" motion of the electron. When Schrodinger's equation is separated, it is found that the radial part of the wavefunction, R, must be a solution of the quantized differential equation $\frac{\hbar^2}{2m}\frac{1}{r^2}\frac{d}{dr}\left(r^2\frac{dR}{dr}\right) + \left(E - \frac{l(l+1)\hbar^2}{2mr^2} + \frac{e^2}{4\pi\varepsilon r}\right)R = 0$
This can be expressed in a simpler form by defining the function
$$f = rR$$
which is then found to satisfy the equation $\frac{\hbar^2}{2m}\frac{d^2 f}{dr^2} + (E - V_{eff})f = 0$
where the effective potential is given by $V_{eff} = \frac{l(l+1)\hbar^2}{2mr^2} - \frac{e^2}{4\pi\varepsilon r}$
where the 1's are the quantum orbital numbers. For s orbitals equal to 1=0, the effective potential is just the electrostatic potential of the nucleus. For 1 greater than zero, the first term is equal to the kinetic energy owing to the angular motion of the electron at a distance r and with angular momentum $\sqrt{l(l+1)}h$.

- [0007]

Referring to FIG. 2, the effective potential is graphed as a function of radius r and the orbital numbers. For orbital number l=0, which is the lower of the three curves, the potential does not provide a stable position for the electron (black disk) and the negatively-charged electron just crashes into the positively-charged proton nucleus as shown by the arrow. For orbital l=1, the first term is called the centrifugal repulsion which together with the electrostatic potential provides for a stable position for the electron as seen in the middle curve. The potential energy is negative which creates a slight valley in which the electron obtains a stable orbit. Higher orbital numbers l=2 produce a similar valley potential further out on the radius. The l=1 orbital does not allow the electron to come near the proton which provides for a stable hydrogen atom. So the key to this invention is how to destabilize this hydrogen atom energy system and produce usable energy which can propel the electromagnetic spacecraft and run other hyperspace inventions. Refer to my patent applications such as Dipole Moment Spacecraft, Dual Potential Hull Spacecraft, Photon Spacecraft, Electromagnetic Field Propulsion System, Full Body Teleportation, Magnetic Vortex Wormhole Generator, Electric Vortex Wormhole Generator, Sulfur S8 Wormhole Generator, Cavitating Oil Hyperspace Energy Generator, Rotor Inductance Propulsion System and Triangular Spacecraft.

- [0008]

Many of these patent applications involve wormholes and hyperspace which are not well-known concepts in the scientific community. Hyperspace consists of those dimensions which are co-dimensional with our spacetime. The reason I know about hyperspace is because (1) I have been in hyperspace on a number of occasions and have experienced Einstein's time dilation according to his General Theory of Relativity, (2) I have experienced more than one full-body hyperspace teleportation over a distance of 100 meters, (3) we have been able to create a wormhole between space and hyperspace with the magnetic vortex wormhole generator in which smoke was blown through one side of the coil into hyperspace, a first contact verified by the Grey Aliens, (4) I have seen the green mist associated with moving out of dimension and crossing over into hyperspace, (5) I have looked into another dimension and have seen another building, a car and a man who waving at me in the presence of an artificially-intelligent Cyborg with the "high-tech look" from the Pleiadian Defense Department, (6) I can remote view through hyperspace subgeometry to distances of 100,000 light years to the edge of the galaxy and have made first contact with around 500 extraterrestrial civilizations involving the use of patent applications such as Remote Viewing Amplifier, Quantum Dot Energy Cylinder and Walking Through Walls Training System, (7) I am the only person on Earth who has communicated with the designers of the crop circles found in England and explained their design to them in terms of subspace geometry, (8) I can walk through walls as a hyperspace energy being, and (9) I have discovered how anti-gravity is possible using low density hyperspace energy, not to mention all the other research work on these electromagnetic field propulsion vehicles. For my work in developing the geometry of the subspace manifold known as the tetrahedron diagram, I was given the Aphysics award by the Admiral, who is third in line to the Admiralty of the Pleiadian Defense Department. For my work in Revelations, she awarded me four beautiful galloping white riderless horses of the Apocalypse. The reason that the Pleiadian Defense Department was involved was that the Admiral had the task of creating the energy being that would protect the subspace manifold during the battle of Revelations which took place in the year 2001. So these are some of my personal experiences in the field of hyperspace physics.

- [0009]

As to the scientific aspect of hyperspace physics, observations of the solar system have noted that large vortices occur on the planets at a latitude of ±19.5°. On the planet Jupiter, for example, the Giant Red Spot vortex, which is the size of two planets like Earth, is located at −19.5° latitude.

- [0010]

Referring to FIG. 3, the Olympus Mons volcano, which is the size of France, occurs in the northern hemisphere at 19.5°. The plume of volcanic ash can be seen being blown to the upper right.

- [0011]

Here on Earth, just north of me at +19.5° in the Caribbean, there is a slow moving rock mantle vortex that curves the islands down toward Venezuela. Since the tetrahedron has three corners, another corner occurs at 120° to the west where the Marshall Islands are located. The reason that the volcanic islands form is that low density hyperspace energy softens the atoms and molecules to such an extent that the atomic bonds are broken. The hot magma from the core has an easier path to the surface through the soft rock compared to the hard rock of surrounding areas. As the vortex rotates, the rising hot magma creates volcanoes which form the chain of islands.

- [0012]

As another example of this, the Silver bridge, which is located at a double harmonic of 39° between Ohio and West Virginia, collapsed because a large wormhole opened up and flooded the bridge with low density hyperspace energy. The metal rivets softened and pulled out of the plates holding down the suspension cables. Thus the roadway tilted to one side and all the cars fell into the river.

- [0013]

If you look at the quarried granite polygonal megalithic stone blocks that were used to build Machu Picchu, the blocks are put together as though they were soft putty. They fit together perfectly. Machu Picchu, or Great Picchu, is the Quichua name for a sharp peak which rises ten thousand feet above the sea. How this was achieved is described in my patent application *Rotating Magnetic Vortex Generator* which shows that rotating permanent magnets can create a wormhole between space and hyperspace. The low density hyperspace energy then floods the block as to make it weightless. The block is then floated up from the quarry which is 2500 feet below Machu Picchu. The huge 1000-ton blocks found in the platform at Baalbek in the Cedar Mountains of Lebanon were also lifted into position in this manner.

- [0014]

Using a technique called Chi Kung breathing, not necessarily known to the Chinese, one of our researchers has been able to levitate himself to a height of six feet. He might have gone higher but he lost his balance and his arm touched a nearby metal pipe. He then floated back to the ground. Contracting the abdominal muscles causes hot air to rise in the lungs while at the same time cold air is inhaled through the nose. This combination of hot and cold air mixing together creates environmental oscillators having a temperature fluctuation. This changes Planck's constant such that the body goes out of dimension and low density hyperspace energy fills the body.

- [0015]

The final example comes from Biblical times. Jesus and his father Joseph were both carpenters. A chair that Joseph had made for a client was found to be too small. So Jesus placed his hands on either side of the chair and stretched the chair to fit. The phrase in Spanish is "mas silla" which means that Jesus made more chair. In the dictionary, the Spanish word for putty is "masilla." The reason that Jesus could do this is that he passed hyperspace energy through his hand vortices which softened the wood. See my patent application called *Hyperspace Torque Generator* which shows this in more detail. The hand vortex is actually a planar co-gravitational K field that can rotate a pendulum around in circles. This solenoidal field corresponds to the magnetic H field. The electric field corresponds to the linear gravitational g field. The equations are identical for both sets of fields. I learned that Jesus actually reincarnated here on Earth in order to identify the planet on which the battle of Revelations would take place.

- [0016]

In summary, these are some examples as to how low density hyperspace energy can soften and break the bonds of atoms and molecules.

- [0017]

As mentioned previously, the astronomical observations suggest that the structure of the universe is related to the tetrahedron. Referring to FIG. 4, the corners of a tetrahedron circumscribed by a sphere, touch the sphere at an angle of 19.47122063°. The ratio of the area-to-volume ratio of the sphere to the area-to-volume ratio of the tetrahedron is ⅓ which is the magic ratio in physics. The arcsin(⅓) is 19.47°.

- [0018]

In order for all the physics constants to be projected into our universe and the co-dimensional hyperspace dimensions, there has to be a tetrahedral subspace manifold. The only mathematical function which allows multiple rotations of the complex plane is the logarithmic function Log[z].

- [0019]

Referring to FIG. 5, the complex number z in the complex plane is equal to x+I y where x and y are real constants and $I^2=-1$. The complex number z can also be written in terms of a radius and an angle
$z=r \, Exp[I \, \theta]$
Because the angle can rotate more than once in 2π m multiples, the complex number z is actually
$z=r \, Exp[I(\theta+2\pi m)]$
Taking the logarithm of z
$Log[z]=Log[r]+I(\theta+2\pi m)$

- [0020]

Referring to FIG. 6, it can be seen that it is not possible to go around multiple times on the same surface due to a branch cut along the origin where Log[0] is undefined. What happens is that as you rotate around on surface (4) and get to the origin, you go down along the branch (5) to the next hyperspace plane (6). Thus the universe is composed of many hyperspace co-dimensions. From personal experience, I estimate that the distance separating the two dimensions is about 3 meters when I was looking into another dimension at the man waving at me. The physics is even more complicated because there is a Lorentz dimensional transformation in which higher dimensions appear smaller and lower dimensions appear larger. In one case I was looking at a huge mothership at a much higher dimension and it looked like a tiny toy model spacecraft. They fired a laser cannon at me, and I then curved space which made the beam change course. You can see why I was awarded the four horses.

- [0021]

    The antilog of Log[z] is
    $$e^{Log[z]} = z = re^{i\omega t}$$
    where the angle is the angular frequency ω times the time t. The subspace geometry remains stationary or fixed, but the projection of the antilog into our dimension generates a system which is frequency dependent. This is why we get oscillating fields.

- [0022]

    From Einstein's General Theory of Relativity, it is known that various kinds of energy can curve spacetime such as mass, electromagnetic fields, angular momentum and electrical charge. The elemental spacetime length ds in cylindrical coordinates {t,r,θ,z}, known as the Schwarzschild metric, shows that spacetime can be curved using mass M and charge Q $(ds)2 = -(dt)2 ? (1 - 2?Mr + Q2r2) + dr2(1 - 2?Mr + Q2r2) + (r? ? d? ? ? θ)2 + (dz)2$
    where you can see that the elemental time dt is dilated by the mass M and the elementary radius dr is reduced. I have actually experienced this time dilation in which, as I was jumping into hyperspace, I was hit by a car which broke my shoulder. When I came back into dimension which appeared to be a few seconds later, I found that I had an 8-inch blood ring down the right side of my chest. My shoulder blade, which sticks up about two inches, is still broken to this day.

- [0023]

    Referring to FIG. 7, flat space (7) can be curved (9) by a massive body (8) such as a planet. For example, the mass of the Earth gives space a negative curvature such that objects tend to fall toward the center of the mass (bowl). On the other hand, using electromagnetic fields, it is possible to produce a positive curvature such that a spacecraft rises by falling upward.

- [0024]

Referring to FIG. 8, if the fields are intense enough, then a wormhole (**12**) forms between space (**10**) and hyperspace (**11**). This depiction is called an embedding diagram because there is no open space going through the wormhole. An object traversing the wormhole moves along the surface from one dimension to another. In order to open the throat (**12**) of the wormhole, negative energy is required. Referring to my patent application Dual Potential Hull Spacecraft, negative energy is produced by the interaction of a microwave beam with an oscillating magnetic H field. In terms of differential forms mathematics, this is given as the Hodge Star * of the differential d of the wedge ^ product of the magnetic H field with the electromagnetic wave {B,E} * d ⊞ (H ^ (B + E ^ dt)) = ∂ (-ρ) ∂ t which says that there is an increasing rate of change of negative energy (−ρ). Due the presence of negative energy together with the spacetime curvature pressure produced by the electromagnetic fields, wormholes open up between space and hyperspace. There is a positive gravitational potential between hyperspace and space because the low density hyperspace energy is more dense than the negative energy in this dimension. Thus the hyperspace energy flows into our dimension which reduces the mass of the spacecraft. The upward spacetime pressure stress over the hull due to the electromagnetic fields creates a lift force on the nearly massless vehicle. Because the hyperspace energy has a speed of light equal to one meter per second, the electromagnetic fields become relativistically strong since they obey the Lorentz transformation. Due to the low mass, high spacetime lift pressure and strong EM fields, the vehicle can attain very high rates of acceleration.

- [0025]

Another method to produce a wormhole is to use bucking magnetic fields which is described in my patent application *Magnetic Vortex Wormhole Generator*. In gravitation physics, the Faraday F tensor, which is a 4×4 spacetime metric {t,x,y,z}, contains all the components of the electromagnetic fields in the various spatial directions {x,y,z} F β α = t x y z ⊡ ▢ 0 E x E y E z E x 0 B z - B y E y - B z 0 B x E z B y - B x 0 ▢ where contravariant index a refers to the rows, and the covariant index β refers to the columns. For example, the component $F^t_x = E_x$ is the electric field in the x-direction. If there were two magnetic bucking fields in the x-direction the Faraday tensor would be F β α = t x y z ⊡ ▢ 0 0 0 0 0 0 0 0 0 0 B x - B x 0 0 - (B x - B x) 0 ▢ = ▢ 0 0 0 0 0 0 0 0 0 0 0 0 0 0 0 0 ▢ which says that the Faraday tensor is zero. Thus no spacetime curvature is generated from two bucking magnetic fields at the same position. On the other hand, if the two bucking magnetic fields are concentric at different radii, then the Faraday tensor becomes F β α = t x y z ⊡ ▢ 0 0 0 0 0 0 0 0 0 0 B x ⊡ δ ⊞ (x 1) - B x ⊡ δ ⊞ (x 2) 0 0 - (B x ⊡ δ ⊞ (x 1) - B x ⊡ δ ⊞ (x 2)) 0 ▢ which is not zero due to the presence of the Kronecker δ delta function which locates the fields at different positions. The spacetime stress-energy-momentum tensor T is then computed from the following equation 4 ▢ ▢ ▢ π ▢ ▢ ▢ T μ ▢ ▢ ▢ v = F μ ▢ ▢ ▢ α ▢ F α v - 1 4 ▢ g μ ▢ ▢ ▢ v ▢ F α ▢ ▢ ▢ β ▢ F α ▢ ▢ ▢ β where g is the metric tensor containing the coefficients of the elemental spacetime length ds. With the mass M and charge Q term equal to zero, there being just electromagnetic fields involved, the g metric tensor in cylindrical coordinates becomes g α ▢ ▢ β = t r θ z ▢ ▢ ▢ - (1 - 2 ▢ M r + Q 2 r 2) 0 0 0 0 (1 - 2 ▢ M r + Q 2 r 2) - 1 0 0 0 0 r 2 0 0 0 0 1 ▢ = ▢ - 1 0 0 0 0 1 0 0 0 0 r 2 0 0 0 0 1 ▢

- [0026]

Referring to FIG. 9, the magnetic vortex generator has two concentric, thin, flat cylindrical silicon-iron cores (**13,14**), each consisting of a stack of three 0.020 inch thick transformer laminations wrapped with insulating tape. Using insulated magnetic wire, a flat helical coil (**16**) is wrapped counter-clockwise around the outer laminations. The coil receives power through connection (**15**). The winding then continues around the core until coil (**17**) where the wire (**18**) is extended to the inner core and the wrapping of coil (**19**) proceeds in the clockwise direction. The inner core wrapping terminates on coil (**20**) with the extension of the second power connection (**21**). Because the coils are wound in opposite directions, the generator produces two bucking magnetic fields at different radii (**22,23**).

- [0027]

According to Maxwell's electromagnetic equations, the curl of the magnetic flux density B field times the square of the speed of light squared is equal to the rate of change of the electric E field $c^2 \nabla \times B = \partial E \partial t$

Multiplying both sides by the elemental area of the core and integrating over the area $c^2 \int_0^{r_0} \nabla \times B \, dA = \partial \partial t \int_0^{r_0} E \cdot n \, dA = \partial \partial t \int_0^{r_0} E \cdot n \, 2 \pi r \, dr$

The curl of the field can be converted into a line integral around the core contour using Stokes' Theorem $c^2 \int_0^{r_0} \nabla \times B \, dA = c^2 \oint_{r_0} B \cdot ds = c^2 \int_0^{2\pi} B \, r_0 \, d\theta$

- [0028]

The magnetic flux density B field oscillates with angular frequency ω
$B = B_0 e^{i\omega t}$

Substituting this into the equation and integrating over time for the inner core field - $c^2 \int_0^t \int_0^{2\pi} B_0 \, e^{i\omega t} \, r_0 \, d\theta = \int_0^{r_0} E2 \, \pi \, r \, dr = E \, \pi \, r_0^2$

This can be solved for the electric E field that is produced by the oscillating magnetic flux density B field $E_0 = 2 I B_0 c^2 (-1 + e^{i\omega t}) r_0 \omega$

The electric field for the outer core is the same equation except that the opposite sign of the magnetic flux density B field and the radius r1 $E_1 = -2 I B_0 c^2 (-1 + e^{i\omega t}) r_1 \omega$

Adding these two fields together is the total electric field E $E = E_0 + E_1 = -2 B_0 c^2 (-1 + Exp(I \omega t)) (r_0 - r_1) r_0 r_1 \omega$

The electrostatic energy of the field is equal to half the linear capacitance of space times the summation over the volume of the dot product of the electric field with itself $U = \varepsilon 2 \int E \cdot E \, dV$

Because the electric field points in the z-direction out of the coil, the dot product is actually the square of the electric field.

- [0029]

Referring to FIG. 10, the energy U per volume is plotted as a function of time with a radius ratio r_1/r_0 of 3/1. As the graph shows, the magnetic vortex wormhole generator produces mostly negative energy which is required in order to create the wormholes. Because the stress-energy-momentum T tensor is also the square of the electric field, this graph gives the spacetime curvature pressure. Thus the electric field produces both the pressure and negative energy required to open up wormholes between space and hyperspace.

- [0030]

Going back to the effective potential equation, V eff = l ⃞ (l + 1) ⃞ ℏ 2 2 ⃞ mr 2 -
e 2 4 ⃞ ⃞ π ⃞ ⃞ ⃞ ε ⃞ ⃞ ⃞ r
it can be seen that the first term is divided by the mass of the particle. In current gravitation physics, the mass of the particle is invariant with velocity. It does not obey the Lorentz transformation. The mass is related to the energy E of the particle and its momentum p by
$m^2 = E^2 - p^2$
In different inertial frames moving with a relative velocity v, the energy and the momentum obey the Lorentz transformation, but no matter what the relative motion, the mass of the particle is constant. In the first term of the effective potential, the mass is constant which leaves just Planck's constant.

- [0031]

  Having worked for over ten years on the subspace manifold, known as the tetrahedron diagram, I found a most incredible intersection on the diagram when working with the water molecule. The water molecule has two hydrogen atoms and one oxygen atom as mentioned previously. The atomic weight of one atom of hydrogen is 1.008 atomic weight units (awu). The atomic weight of oxygen is 16.000 awu. Therefore the molecular weight of water is Weight ⃞ ⃞ ⃞ of ⃞ ⃞ ⃞ two ⃞ ⃞ ⃞ atoms ⃞ ⃞ ⃞ of ⃞ ⃞ ⃞ hydrogen ⃞ ⃞ ⃞ 2 × 1.008 ⃞ ⃞ ⃞ awu = 2.016 ⃞ ⃞ ⃞ awu Weight ⃞ ⃞ ⃞ of ⃞ ⃞ ⃞ one ⃞ ⃞ ⃞ atom ⃞ ⃞ ⃞ of ⃞ ⃞ ⃞ oxygen ⃞ ⃞ ⃞ 1 × 16.000 ⃞ ⃞ ⃞ awu = 16.000 ⃞ ⃞ ⃞ awu 18.016 ⃞ ⃞ ⃞ awu
  The gram molecular weight is the atomic weight expressed in grams, so there are 18.016 grams in Avogadro's number of molecules. So the mass per molecule in logs is Log [18.016 ⃞ ⃞ ⃞ gram - mol / (1000 ⃞ ⃞ ⃞ gm kg) 6.02 × 10 23 ⃞ ⃞ ⃞ mol] = - 58.77103943

- [0032]

  Referring to FIG. 11, the tetrahedron diagram plots the natural logarithm of mass versus the natural logarithm of wavelength. The reason for this is that mass times wavelength is equal to Planck's constant divided by the speed of light c, known on the diagram as the base constant. At the present time there are over 4000 diagrams which are copyrighted in the Library of Congress. In logarithms, the product of two numbers is the sum of the two numbers. This means that the sum of the mass and the wavelength are equal to the base constant which has a value of
  ln[m]+ln[λ]=ln[h/c]=−95.91546344=base constant
  Our dimension has a lower limit on mass and length known respectively as the Planck mass and the Planck wavelength. The Planck mass is the linear mass Ω of the universe times the Planck scale Λ. The Planck wavelength is circumference of a circle of radius Planck scale. In terms of logs, the Planck mass and Planck wavelength are
  Planck mass=ln(ΩΛ)=−17.64290101
  Planck wavelength=ln(2πΛ)=−78.27256243

- [0033]

  When these values are plotted on the tetrahedron diagram shown in FIG. 11, the Planck box (abcd) is formed which are the boundaries of our dimension in subspace. The line numbering is as follows Planck mass 25, 27 Planck wavelength 26, 28 mass of water molecule 29, 31 speed of light squared circle 30 base constant 33 inverted tetrahedrons 34, 35 centerline 36

The energy of the water molecule, circle (37), is equal to the sum of the water molecule mass (29) plus the speed of light squared circle (30). The energy circle (37) intersects the mass of the water molecule (29) at the Planck wavelength (28, point e), which is the boundary between space and hyperspace. What this means is that the mass is equal to the energy at the Planck box boundary. The only way that this is possible is if the speed of light c is equal to one meter per second
$E=m c^2 = m c = 1$ meter/second
A water molecule traversing a wormhole into hyperspace undergoes a change in the speed of light from 299792458 m/s to 1 m/s.

- [0034]

  Planck's constant $\hbar$ is equal to the Planck mass $\Omega\Lambda$ times the Planck scale Λ times the speed of light c.
  $\hbar = \Omega\Lambda\Lambda c$
  By having the speed of light go to 1 m/s, the orbital term in the effective potential V_{eff} is reduced by a factor of the speed of light squared equal to 9×10^{16}. This unbalances the equation to such an extent that only the Coulomb potential term remains. l ⬚ (l + 1) ⬚ $\hbar$ 2 2 ⬚ m ⬚ ⬚ ⬚ r 2 ⬚ coulomb ⬚ ⬚ ⬚ term
  The electron is attracted to the proton nucleus because the centrifugal term no longer provides a stable orbit for the electron. Thus the atomic binding is destroyed and the water molecule becomes soft as putty.

- [0035]

  Referring to FIG. 12, the collision of the electron with the proton together with the enormous change in the proton's energy causes the proton p to become unstable and decay. According to the Standard Model of particle physics, the elementary particles are composed of smaller particles known as quarks. The six quarks have been named up u, down d, strange s, charm c, top t, and bottom b. The subscript on the quark indicates one of three colors {red r, blue b, green g}. As shown in the diagram, the proton is composed of three quarks {u_r, u_g, d_b}, two of which are up quarks of which one is red and the other green, and a third blue down quark. The proton p decays into a positron e⁺ which is an electron with a positive charge, and a neutrally-charged pion π^0 particle through the exchange of an X boson particle. The pion has a mass between the electron and the proton.
  $p \rightarrow \pi^0 + e^+$

- [0036]

  Referring to FIG. 13, the pion π^0 then decays into a proton p and antiproton {overscore (p)} which annihilate each other to produce two photons shown on the right by the traveling waves. So the overall energy exchange is p $\rightarrow$ 2 ⬚ ⬚
  ⬚ hv c 2 + e +
  where hv is the energy per gamma photon with frequency v. The electron of the hydrogen atom would then annihilate the positron for additional photon energy.

- [0037]

  Referring to FIG. 14, the hydrogen H atom is composed (38) of the proton and electron as seen in the upper left corner. The proton decays (39) into the neutral pion and a positron. The electron from the hydrogen atom and this positron form one electron pair (40). The pion then decays (41) into two gamma photons which produce an electron pair production energy cascade into 132 pairs (42,43) for a total of 133 electron pairs. These electrons can then be captured electrostatically and used for the production of electricity.

- [0038]

  Referring to FIG. 15, the electrons are captured with the water droplet injector . The plunger (46) of a spring-loaded cylindrical solenoid (44) is attached to a tapered piston (47). By means of ring collar and bolts (45), the solenoid is bolted to the injector (48). A supply of purified water is attached to the water inlet connection (49). When the solenoid is activated, it pulls back slightly so that water can enter the valve. When the solenoid is deactivated, the piston forces the water droplet out through the nozzle (50) into a cylindrical glass vacuum chamber (52). Two cylindrical glass disks (51) hold the nozzle in place. On the other end of the vacuum chamber is the sealed-tube connection (56,57,58) to the vacuum pump. In the middle of the vacuum chamber, two metal plates (53) are attached through sealed glass collars (54) to electrical pins (55). The plates are electrostatically-charged with opposite charges so as to form a capacitor. This creates an electrical field between the plates which attracts the electrons to the positively-charged plate.

- [0039]

  Referring to FIG. 16, the vacuum tube and water injector (61) are mounted along the centerline of the inner (60) and outer (59) magnetic vortex wormhole generator coils. The low density hyperspace energy traversing the wormhole along the centerline of the coils causes the injected water molecules to soften and decay into a cascade of electrons. The oscillating electric field along the centerline causes the electrons to vibrate back and forth. The crossed electric field between the charged capacitor plates causes the electrons to flow toward the positively-charge plate in order to produce electricity.

- [0040]

  Referring to FIG. 17, the vacuum tube is connected to the vacuum pump through a hose connection to the pump air inlet (64). A 5 Hp electric motor (62) drives dual rotating flights of screws which trap the air and move it toward the exhaust outlet (65) shown with no muffler. On a spacecraft operating in the vacuum of outer space, this component would not be needed.

  SUMMARY OF THE INVENTION

- [0041]

  It is the object of this invention to generate electricity by using low density hyperspace energy to soften water molecules such that the atomic binding is broken which causes the hydrogen nucleus to decay into a cascade of electron pairs. These electrons are then collected on a positively-charged plate in order to produce electricity. The water molecules are softened by flooding them with low density hyperspace energy that is produced by a magnetic vortex wormhole generator. The generator creates negative energy and a spacetime curvature along the centerline of two concentric coils. This combination opens up wormholes along the centerline. Because the gravitational potential of low density hyperspace energy is greater than the negative energy, the hyperspace energy flows through the wormhole from hyperspace into our dimension. The hyperspace energy has a speed of light equal to one meter/second. This causes a change in Planck's constant h such that the proton orbitals of the hydrogen atom are unable to produce a centrifugal repulsion which keeps the electron in orbit. The Coulomb potential term dominates and the electron is attracted to the proton. Due to the vast change in the speed of light, and the collision of the electron with the proton, the proton becomes unstable and decays into a neutral pion and a positron. The pion then decays into two gamma photons which produce a large cascade of electron pairs.

- [0042]

  A water injector, consisting of a solenoid-activated valve and nozzle, injects water droplets into a vacuum chamber which is positioned along the centerline of the two concentric coils where the wormholes form. Due to the low density hyperspace energy passing through the wormholes into our dimension, the water molecules soften and decay into electrons which are collected on an electrostatically-charged capacitor plate having a positive charge located in the glass vacuum chamber.

  A BRIEF DESCRIPTION OF THE DRAWINGS

- [0043]

  FIG. 1. Perspective view of hydrogen atom K shell.

- [0044]

  FIG. 2. Graph showing potential binding energy of hydrogen atom.

- [0045]

  FIG. 3. Perceptive view of Olympus Mons volcano at +19.5° Mars latitude.

- [0046]

  FIG. 4. Perspective view of tetrahedron inscribed in sphere.

- [0047]

  FIG. 5. Graph showing complex plane.

- [0048]

  FIG. 6. Perspective view of hyperspace co-dimensions of logarithmic manifold.

- [0049]

  FIG. 7. Perspective view of embedding diagram showing curvature of space caused by a mass.

- [0050]

  FIG. 8. Perspective view of wormhole embedding diagram.

- [0051]

  FIG. 9. Perspective view of coils of magnetic vortex wormhole generator.

- [0052]

  FIG. 10. Graph showing that generator produces negative energy.

- [0053]

FIG. 11. Tetrahedron diagram showing that the speed of light at the Planck box boundary at the water molecule is one meter/second.

- [0054]

FIG. 12. Perspective view of proton decay into neutral pion and positron.

- [0055]

FIG. 13. Perspective view of pion decaying into two gamma photons.

- [0056]

FIG. 14. Diagram showing decay of the hydrogen atom into electrons.

- [0057]

FIG. 15. Perspective view of water injector and vacuum chamber.

- [0058]

FIG. 16. Perspective view of vacuum chamber mounted along centerline of magnetic vortex wormhole generator.

- [0059]

FIG. 17. Perspective view of vacuum pump used to evacuate vacuum chamber.

DETAILED DESCRIPTION OF THE INVENTION

- [0060]

1. The coils of the magnetic vortex wormhole generator are made of three stacks of 0.020 inch silicon-iron transformer laminates. These are washed to remove the oil, and then wrapped with insulating tape in order to keep the laminations together. Using a very long bench made of wooden planks and 2×4 sawhorses, the outer coil is wrapped counter-clockwise right to left using a large spool of 14 AWG magnetic wire. A thin spacer is used between windings in order to reduce the winding capacitance. Once the outer coil is wound, the wire is continued to the second inner coil which is wrapped clockwise, leaving enough wire between coils such that when the coils are mounted in the wooden frame, the coil is one continuous winding having an input and output connection. Using an inductance meter, the inductance of the coil is measured. Using a standard frequency of 60 Hz, the capacitance of a sheet metal capacitor is calculated such that the generator is electromagnetically resonant at this frequency. The generator is connected to the line voltage by a 1:1 isolation transformer which is connected to a small primary coil wrapped on a toroidal core whose similar secondary coil is connected to the sheet metal capacitor and inductance coil. Resonance is achieved by adjusting the spacing and overlap of the sheet metal.

- [0061]

2. The vacuum chamber is made of a glass tube with sufficient wall thickness to withstand the vacuum pressure. A number of glass blowing techniques are used to make the glass-electrode connection for the capacitor plates. Then circular pieces of glass plate are cut out and ground to the inside diameter of the tube, fitted with the nozzle and vacuum connection, and then heat sealed to the chamber. The chamber and water injector are then attached to a wooden bracket mounting which is doweled and glued to the wooden frame of the generator.

Claims

1. A water energy generator system comprising the components:
a magnetic vortex wormhole generator and driving resonant electrical circuit;
a water droplet injector;
a vacuum chamber and vacuum pump; and
an electrostatic electron capture system.

2. By means of claim (**1**), a magnetic vortex wormhole generator comprising two concentric cylindrical coils of different radii wound in opposite directions, made of thin sheet silicon-iron transformer laminations wound with one continuous length of magnetic wire providing a single input and single output connection to the driving electrical circuit.

3. By means of claim (**2**), a coil winding method and oscillating driving circuit producing bucking electric fields along the centerline of the generator normal to the coils which create a spacetime curvature pressure and negative energy.

4. By means of claim (**3**), the generation of wormholes between space and hyperspace along the centerline of the generator such that low density, low speed of light hyperspace energy flows through a positive gravitational gradient from hyperspace to space.

5. By means of claim (**1**), a water injector comprising a solenoid-activated water valve, water supply connection, seal and nozzle for injecting water droplets into the vacuum chamber.

6. By means of claims (**5**) and (**4**), the softening and particle decay of the water molecules by the hyperspace energy into neutral pions, positrons, gamma photons and finally a cascade of electron pairs.

7. By means of claims (6) and (1), the capture of the electrons on electrostatically charged capacitor plates located in the vacuum chamber for the purpose of producing electrical energy.

8. By means of claims (1) and (2), a resonant electrical driving circuit comprising a line isolation transformer connected to a primary coil wound on a toroidal coil whose secondary output coil is connected to a capacitor and the inductance coil of the generator such that the capacitance of the capacitor and the inductance of the coil form a highly resonant electrical circuit.

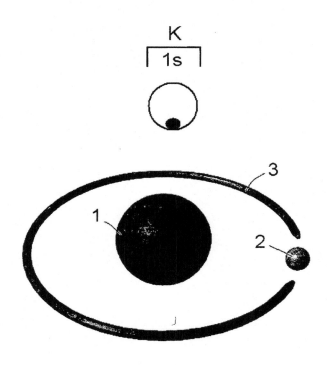

Figure 1

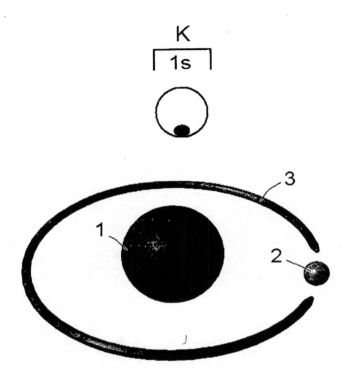

Figure 2

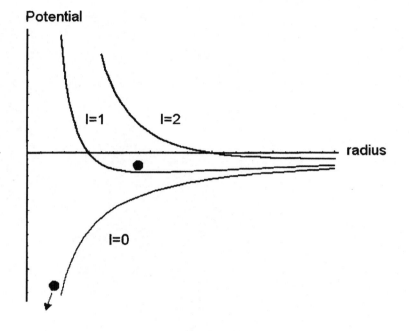

Potential

l=1   l=2

radius

l=0

Figure 3

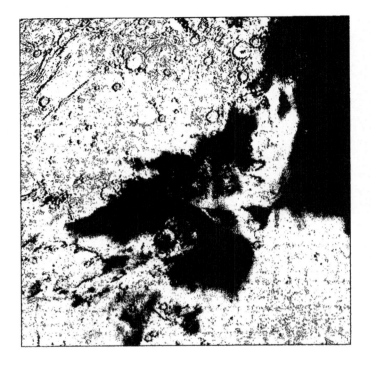

Figure 4

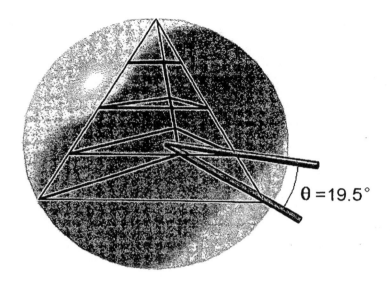

$\theta = 19.5°$

Figure 5

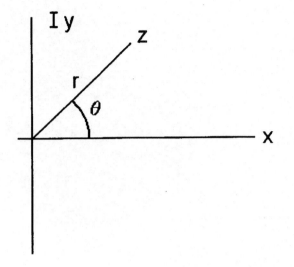

Figure 6

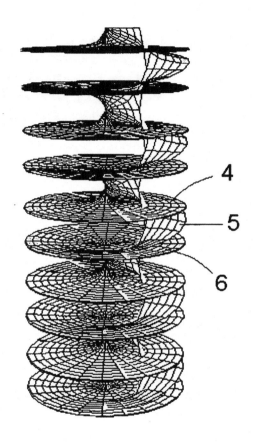

Figure 7

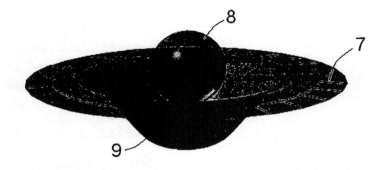

Figure 8

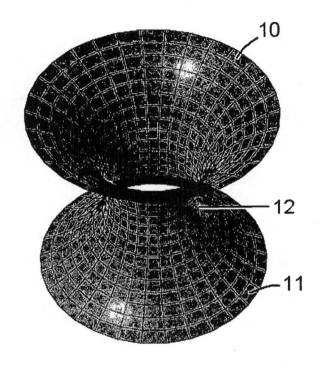

Figure 9

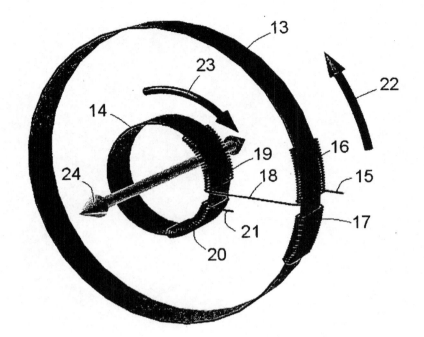

Figure 10

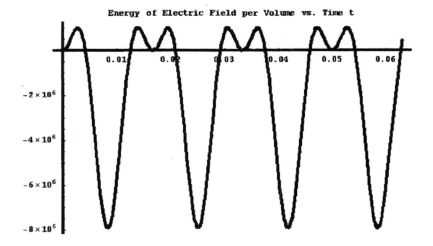

Figure 11

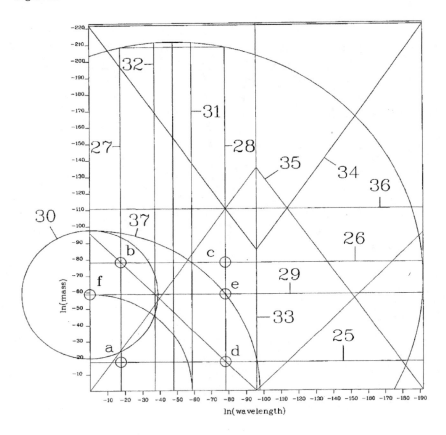

Figure 12

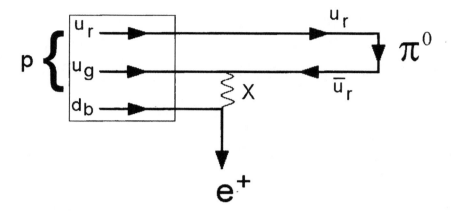

Figure 13

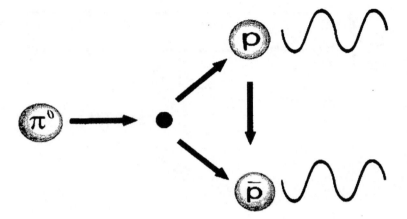

Figure 14

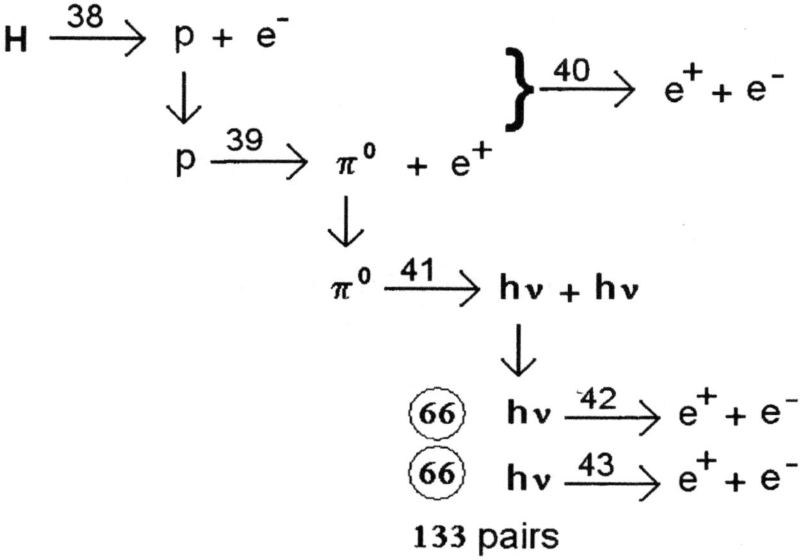

$$H \xrightarrow{\ 38\ } p + e^-$$

$$\downarrow$$

$$p \xrightarrow{\ 39\ } \pi^0 + e^+$$

$$\left.\right\} \xrightarrow{\ 40\ } e^+ + e^-$$

$$\downarrow$$

$$\pi^0 \xrightarrow{\ 41\ } h\nu + h\nu$$

$$\downarrow$$

$$\text{(66)} \quad h\nu \xrightarrow{\ 42\ } e^+ + e^-$$

$$\text{(66)} \quad h\nu \xrightarrow{\ 43\ } e^+ + e^-$$

133 pairs

Figure 15

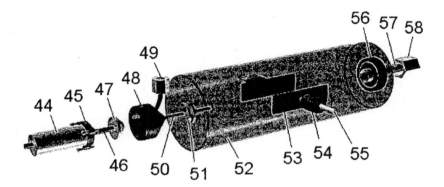

Figure 16

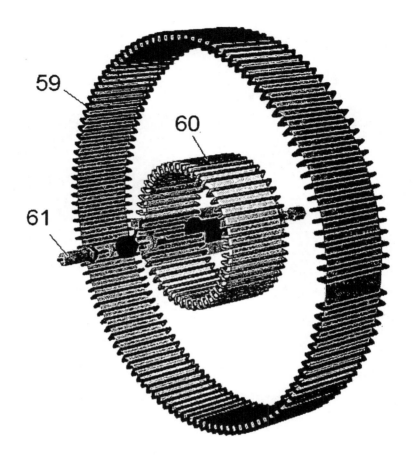

Figure 17

62

63

64

65

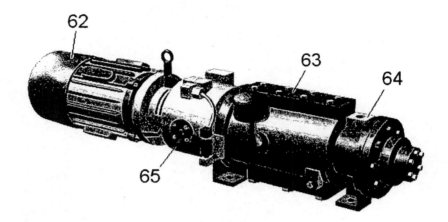

The End?